Basic Control Volume Finite Element Methods for Fluids and Solids

IISc Research Monographs Series

Basic Control Volume Finite Element Methods for Fluids and Solids

Vaughan R Voller

University of Minnesota, USA

NEW JERSEY • LONDON • SINGAPORE • BEIJING • SHANGHAI • HONG KONG • TAIPEI • CHENNAI

Published by

World Scientific Publishing Co. Pte. Ltd.
5 Toh Tuck Link, Singapore 596224
USA office: 27 Warren Street, Suite 401-402, Hackensack, NJ 07601
UK office: 57 Shelton Street, Covent Garden, London WC2H 9HE

British Library Cataloguing-in-Publication Data
A catalogue record for this book is available from the British Library.

BASIC CONTROL VOLUME FINITE ELEMENT METHODS FOR FLUIDS AND SOLIDS
IISc Research Monographs Series — Vol. 1

ISBN-13 978-981-283-498-0
ISBN-10 981-283-498-2

Printed in Singapore.

IISc Research Monographs Series

World Scientific Publishing Company (WSPC), Singapore and Indian Institute of Science (IISc), Bangalore co-publish a series of state-of-the-art monographs written by experts in specific areas. They include, but are not limited to, the authors' own research work.

This pioneering collaboration aims to contribute significantly in disseminating current Indian scientific understanding worldwide. In addition, the collaboration also proposes to bring the best scientific thoughts and ideas across the world in areas of priority to India through specially designed India editions.

Books in the Series:

V. R. Voller, *Basic Control Volume Finite Element Methods for Fluids and Solids* (2009).

To my mentors, advisors, colleagues, and students

Preface

The advent of the digital computer in the middle of the last century initiated a rapid and continued growth in the development of computational tools for solving field problems. In the early days of these developments, researchers worked on a relatively small set of methods, applying them to a broad range of mechanics problems from stress and strain in solids through to fluid flow. As the field matured, however, distinct camps of researches based around methods and problems were formed. In extremely broad terms, the development of methods were split between those based on finite difference approaches and those based on finite element approaches; likewise applications were split between solids and fluids.

As the computational modeling field moved forward, other classes of solution methods were developed. Of particular note were control volume/finite volume methods. An immediate appeal of such methods was their obvious connection, through explicit discrete balance equations, to the physics of the problem at hand. Early control volume developments used finite difference methods to arrive at appropriate discrete equations. It was rapidly realized, however, that control volume solutions could also be constructed through the use of finite element technologies. Thus, control volume methods are viewed by some researchers as bridging between finite difference and finite element methods, with the ability to adopt and adapt the advantage of these methods while neglecting the drawbacks.

The Control volume methods that seem to obtain the maximum advantage of this hybrid view point are those based on finite element technologies, referred to as Control Volume Finite Element Methods (CVFEM). A notable feature of this class is the relative ease by which they can be applied to both solids and fluids problems. As such, the current research focused on solving multi-physics problems has spurred a significant interest in developing CVFEM solutions.

The central aim of this monograph is to introduce the basic and essential ingredients in control volume finite element methods. It is felt that this introduction will provide researchers with the critical background and base tools that will allow for more general application of CVFEM. Further, looking toward future multi-physics applications and trying to recapture the more comprehensive approach of the early days of computational mechanics, this monograph develops the basic constructions of CVFEM in the context of solving fundamental problems in both solids and fluids. This approach serves to fully highlight the generality and flexibility of CVFEM.

As with all efforts of this nature there is a host of people to thank. In general terms, I would first like to thank all of my mentors, advisors, colleagues, and students who have greatly contributed to my current understanding of computational mechanics. More explicit thanks need to go to the faculty of the Department of Mechanical Engineering at the Indian Institute of Science in Bangalore for providing motivation and support in this effort. I am also indebted to Mr Jim Hambleton who acted as a great sound board during the writing and provided a proof reading of the draft text.

A list of errata can be found at

http://personal.ce.umn.edu/~voller/errata.pdf

Readers are welcome to contact me (volle001@umn.edu) if they have any queries or suggestions.

V.R. Voller
Civil Engineering
University of Minnesota

Contents

Chapter 1

Introduction

A very brief overview to numerical method for solids and fluids is presented. The objectives and philosophy of the work in this monograph are specified. The basic concept in the control volume numerical approach is highlighted. A detailed breakdown of the contents of each chapter is provided.

1.1 Overview

Since the advent of the digital computer in the middle of the 20th century there has been a plethora of numerical methods designed to solve the equations that describe the behavior of solids and fluids. Two popular classes of methods are (i) Finite Difference Methods (FDM) and (ii) Finite Element Methods (FEM). In the former, the problem domain is covered by a grid of node points and the components of the governing equations are approximated using Taylor series expansions. In the latter, the domain is covered by a mesh of elements—geometric shapes defined by nodes at vertices and other strategic locations—and the terms in the governing equations estimated in terms of functions that interpolate the nodal values over the elements. A distinction between the methods is the nature of the nodal locations. In basic approaches, the FDM is restricted to a uniform grid that is constrained to coincide with the coordinate directions. In contrast, the FEM has no such restriction and can operate on an unstructured mesh optimized to fit arbitrary problem domains. As such, it is fair to say that, in solving many practical problems, the relative computational ease and advantage of the FDM loses out to the geometric flexibility of the FEM.

An important variation of the finite difference method that has had extensive application in the solution of fluid flow problems (Patankar (1980)) is the so-called Control Volume Finite Difference Method (CVFDM). In this approach, control volumes are created around the node points on the structured grid. Then, a set of discrete equations is arrived at by appropriate balancing of the control volume boundary fluxes—approximated by Taylor series expansions. An attractive feature of this method is that it has a direct connection to the physics of the system. This is seen by noting that the starting point in the derivation of the governing equations of solid and fluids is the balance between surface fluxes and volumes rates of change over a control volume. Despite this physical attribute, however, the CVFDM is still subject to the geometric constraints of the basic FDM. Starting with the pioneering work of Winslow (1966) the Control Volume Finite Element Method (CVFEM)—sometimes called the Finite Volume Method (FVM)—was developed to overcome this drawback. The key feature to recognize is that control volumes can also be constructed around the node points on an unstructured finite element mesh that conforms to an arbitrarily shaped domain. With this construction, the fluxes across control volume faces can be approximated by using finite element interpolation. Balancing these fluxes, leads to a physically based representation of the governing equation as a discrete set of equations in terms of mesh nodal values.

The original application of CVFEM by Winslow (1966) was directed at electromagnetic field problems. This was followed by applications in heat transfer and fluid flow problems, Baliga and Patankar (1980), Baliga and Atabaki (2006), and solid mechanics problems, Fryer et al (1991).

1.2 Objective and Philosophy

The objective of this monograph is to introduce a single common framework for the CVFEM solution of both fluid and solid mechanics problems. To emphasize the essential ingredients, discussion is restricted to two-dimensional problems solved by CVFEM utilizing linear elements. This allows for the straightforward provision of the key information required to fully construct working solutions of basic fluid flow and solid mechanics problems. Example problems are based on

1. advection-diffusion equations for scalar transport,
2. plane stress and plane strain treatments for linear elasticity, and
3. the stream-function-vorticity form of the two-dimensional Navier Stokes equations for incompressible Newtonian flow.

In developing CVFEM numerical treatments, the most basic discretization schemes are used (e.g., linear elements and up-winding) and the solution of the resulting algebraic equations in the nodal unknowns is based on crude technologies, e.g., fully explicit time integration and point iterative schemes. In this way, our path toward arriving at a working framework for CVFEM solutions is not seriously detoured by unnecessary detail. The contention is that on establishing a basic framework a reader will be in an ideal position to read the relevant literature and readily incorporate the subtle changes that can and will make the CVFEM solutions more efficient and accurate.

1.3 The Basic Control Volume Concept

Although reinforced numerous times throughout this text it is worthwhile in this opening introduction to provide an illustration of the basic physical concept in a control volume method. To do this, consider the polygonal control volume of Fig. 1.1 placed in a steady incompressible two-dimensional flow of a contaminated fluid. Assuming a unit depth and a known fluid velocity field $\boldsymbol{v} = (v_x, v_y)$, the fluid volume flow-rate out across any one of the faces of this polygon is

$$q_{out} = \int_{face} \boldsymbol{v} \cdot \boldsymbol{n} \, dA \qquad (1.1)$$

where $\boldsymbol{n}$ is the outward pointing normal on the face. Since the flow is incompressible the net flow out of the volume is zero and given by

$$\sum_{face=1}^{5} \int_{face} \boldsymbol{v} \cdot \boldsymbol{n} \, dA = 0 \tag{1.2}$$

If there are contaminant sources and sinks in the domain of the fluid flow, the contaminant concentration $C(x, y)$ will vary throughout the field. At any point in this field, the rate of contaminate transported per unit area by the fluid motion (advection) is $\boldsymbol{v}C$ and the rate of contaminate transported by molecular diffusion (assuming isotropic conditions and Fickian diffusion) is $-\kappa\nabla C$. In this way, the steady state balance (net flow out) of the contaminate over the control volume shown in Fig. 1.1 is given by

$$\sum_{face=1}^{5} \int_{face} \boldsymbol{v}C \cdot \boldsymbol{n} \, dA - \sum_{face=1}^{5} \int_{face} \kappa\nabla C \cdot \boldsymbol{n} \, dA = 0 \tag{1.3}$$

The conceptual heart of a Control Volume based numerical method is to develop a means of approximating the integrals and derivatives in (1.3) so as to reduce it to an algebraic equation; an equation written in terms of the values of C at the discrete node points in the neighborhood of the control volume.

To fully appreciate the concept used in the control volume approach it is of high importance to note that equation (1.3) is an exact expression of the underlying physics of the problem at hand. Although the governing partial differential equation for this advection-diffusion problem is more typically written in the point form

$$\nabla \cdot \boldsymbol{v}C - \nabla \cdot [\kappa\nabla C] = 0, \tag{1.4}$$

as will be emphasized in Chapter 2, the integral form in (1.3) is an equally valid governing equation. Furthermore, as will be exploited in Chapter 5, the integral form of the governing equation provides a clear route toward obtaining the CVFEM discrete equations in terms of the nodal values located at the centers of polygonal control volumes.

1.4 Main Topics Covered

The main topics covered in this monograph are as follows.

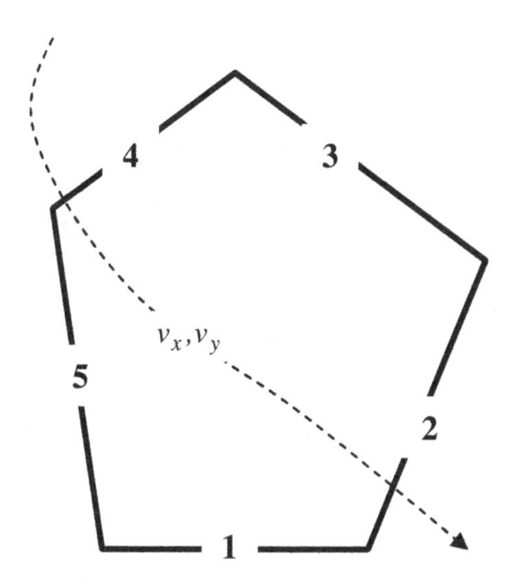

Fig. 1.1 A polygonal control volume in a contaminated fluid flow

In Chapter 2 the governing equations for solid and fluid problems are derived. In order to emphasis the connection between the continuous governing equations and the discrete equations generated by the CVFEM, some care is taken in developing detailed derivations. In particular, the governing equations are developed in both point and integral forms; the latter allowing for a natural connection to the CVFEM discrete equations. The chapter opens by writing down general balance equations for mass, linear momentum, and general scalar quantities. From these general equations, specific governing equations for, advection-diffusion transport, fluid flow, and elasticity problems are derived. This set of equations form the example problems that guide the key developments of the CVFEM.

In Chapter 3 the essential ingredients of a numerical solution in general and a CVFEM solution in particular are presented. The main steps in a numerical solution of a field problem are outlined.

In Chapter 4 a data structure that codifies the critical geometric relationships in an unstructured mesh is developed. A brief outline is provided to show how this data structure can be used to make an automated link between the physics underlying the governing equation and the discrete CVFEM equations.

In Chapter 5, working with a general advection-diffusion equation, detailed derivations of the main components in a CVFEM are presented. This is the essential knowledge kernel in this work.

In Chapter 6, in order to form a contrast, a brief presentation of the development of a Control Volume Finite Difference Method for the advection-diffusion equations is made.

In Chapter 7 the CVFEM of Chapter 5 is fully tested on a comprehensive range of advection diffusion problems. The problems chosen all have analytical solutions that provide meaningful testing of the two-dimensional CVFEM operating on an unstructured grid. The solutions span from steady state diffusion with constant diffusivity through to transient advection-diffusion with variable properties.

In Chapter 8 the CVFEM solution for plane stress and plane strain elasticity is developed. The application of this solution technology to the problem of stress concentration around a hole in an infinite region subjected to a uniform far-field stress is made. Comparisons between the CVFEM and the known analytical solution are provided.

In Chapter 9 the CVFEM solution for the stream-function vorticity form of the two-dimensional Navier Stokes equations for an incompressible Newtonian fluid is developed. CVFEM solutions are compared with results from a high-fidelity numerical benchmark solution.

In Chapter 10 notes toward the developments of a three-dimensional CVFEM are provided. In particular features of tetrahedral elements are noted and the calculation for the flux across a control volume face is presented.

In Appendix A a MATLAB code for generating a triangular mesh with an appropriate CVFEM data structure is provided. Although this mesh is based on a structured grid the resulting data structure can be used with an unstructured CVFEM solution.

In Appendix B a MATLAB code for the CVFEM solution of a steady sate advection diffusion equation is provided.

Chapter 2

Governing Equations

The governing equations for solid and fluid mechanics problems are derived. Examples include advection-diffusion of a scalar, plane strain and plane stress elasticity, and the stream function-vorticity formulation of the Navier-Stokes flow equations.

2.1 The Euler Equations of Motion

The basic equations for describing motion of a deforming continuum Ω—solid or liquid—are the Euler equations, expressing the conservations of mass and linear momentum. These equations can be derived in two ways, (i) performing a balance of mass or momentum on a control volume fixed in space—referred to as the Eulerian approach or (ii) tracking the rate of change of the mass or momentum in a specified volume of the continuum as it moves through space—referred to as the Lagrangian approach. For our purposes of developing Control Volume Finite Element Method (CVFEM) solutions, the former Eulerian description is the most appropriate.

2.1.1 Conservation of mass

In the Eulerian approach mass conservation is expressed as

$$\frac{d}{dt}\int_V \rho dV = -\int_A \rho \boldsymbol{v} \cdot \boldsymbol{n} dA \tag{2.1}$$

where V is a fixed volume in space—arbitrarily chosen from the continuum Ω, A is the surface that encloses V, $\boldsymbol{n}$ is the outward pointing unit normal at a point on this surface, $\boldsymbol{x}$ is the position vector (from a fixed origin) to a point in V, $\rho = \rho(\boldsymbol{x},t)$ is the density (mass/volume) at this point at time t, and $\boldsymbol{v} = \boldsymbol{v}(\boldsymbol{x},t)$ is the material velocity at point $\boldsymbol{x}$ at time t. Equation (2.1) states that the rate of increase of the mass in V is given by the net rate of mass flowing in across its surface A. On taking the time derivative under the integral (volume V is fixed) and using the divergence theorem, (2.1) can be written solely in terms of a volume integral,

$$\int_V \frac{\partial \rho}{\partial t} + \nabla \cdot (\rho \boldsymbol{v}) dV = 0 \tag{2.2}$$

2.1.2 Conservation of linear momentum

In the Eulerian approach, conservation of linear momentum is obtained by applying Newton's second law of motion to the fixed control volume V

$$\frac{d}{dt}\int_V \rho \boldsymbol{v} dV = \int_V \rho \boldsymbol{b} dV + \int_A \boldsymbol{t} dA - \int_A (\boldsymbol{v} \cdot \boldsymbol{n}) \rho \boldsymbol{v} dA \tag{2.3}$$

where $\boldsymbol{b} = \boldsymbol{b}(\boldsymbol{x},t)$ is a body force and $\boldsymbol{t} = \boldsymbol{t}(\boldsymbol{x},\boldsymbol{n},t)$ is the traction, consisting of friction forces acting tangential to the surface A and pressure forces acting normal to the surface A; as indicated in the definition, the traction explicitly depends on the orientation of the boundary over which it acts. The left hand side of (2.3) is the rate of change of the linear momentum in V, which is balanced by the body forces acting in V, the traction applied on the surface A of V, and the net rate of linear momentum flowing into V across A. Through considering the tractions on a vanishing tetrahedron and the angular momentum balance it can be shown that the traction at a point $\boldsymbol{x}$ on area A can be written in terms of a second order symmetric stress tensor $\underline{\underline{\sigma}}$ and the unit outward normal, i.e.,

$$\boldsymbol{t} = \underline{\underline{\sigma}}\boldsymbol{n} \tag{2.4}$$

or in index notation $t_i = \sigma_{ij} n_j$. In this way, on accounting for the fixed nature of V, the momentum conservation in (2.3) can be written as

$$\frac{d}{dt}\int_V \rho \boldsymbol{v} dV = \int_V \rho \boldsymbol{b} dV + \int_A \underline{\underline{\sigma}} \boldsymbol{n} dA - \int_A (\boldsymbol{v}\cdot\boldsymbol{n})\rho \boldsymbol{v} dA \tag{2.5}$$

2.1.3 Conservation of scalar

In addition to conservation of mass and linear momentum we may also need to express conservation of a scalar ϕ [quantity/mass], e.g., chemical composition or heat. A relatively general form of this conservation over the fixed volume V is

$$\frac{d}{dt}\int_V \rho\phi dV = \int_V \rho Q^m dV - \int_A \boldsymbol{q}\cdot\boldsymbol{n} dA - \int_A (\boldsymbol{v}\cdot\boldsymbol{n})\rho\phi dA \tag{2.6}$$

where Q^m is a mass source [quantity/mass-time] and $\boldsymbol{q}$ is a flux vector [quantity/area-time]. In many cases the flux is a diffusion and

$$\boldsymbol{q} = -\underline{\underline{\kappa}}\nabla\Psi \tag{2.7}$$

where $\underline{\underline{\kappa}}$ is a second order symmetric tensor and Ψ is a potential (e.g., temperature).

2.2 Specific Governing Equations

The integral forms of equations (2.1), (2.5) and (2.6) are most suitable for the development of CVFEM solutions. To solve specific problem types, however, it is necessary to define constitutive relationships for the stress tensor $\underline{\underline{\sigma}}$ in (2.5) and flux vector $\boldsymbol{q}$ in (2.6). The end objective in such an exercise is to establish a set of governing equations where the number of unknowns matches the number of equations. Below, restricting attention to a Cartesian coordinate system where $(x_1, x_2, x_3) \equiv (x, y, z)$ and $\boldsymbol{v} = (v_1, v_2, v_3) \equiv (v_x, v_y, v_z)$, examples of the development of specific governing equations are provided.

2.2.1 Mass conservation in an incompressible flow

In an incompressible flow the rate of change of the density of a given mass element following the fluid motion is zero, i.e., in terms of the material derivative

$$\frac{D\rho}{Dt} = \frac{\partial \rho}{\partial t} + \boldsymbol{v} \cdot \nabla \rho = 0 \tag{2.8}$$

Integrating this quantity over the fixed volume V and subtracting the result from both sides of the mass conservation (2.2) results in the following integral forms of mass conservation in an incompressible flow

$$\int_V \rho \nabla \cdot \boldsymbol{v} \, dV = 0 \tag{2.9}$$

or through application of the divergence theorem

$$\int_A \rho \boldsymbol{v} \cdot \boldsymbol{n} dA = 0 \tag{2.10}$$

Since the volume V in (2.9) is arbitrary and the density is positive and non-zero $\rho > 0$, the divergence of the velocity must be identically zero at every point in the domain,

$$\nabla \cdot \boldsymbol{v} = 0, \quad \forall \boldsymbol{x} \in \Omega \tag{2.11}$$

i.e., a divergence-free velocity field is a point-form statement of mass conservation in an incompressible fluid.

Note that, throughout this work, in addition to the rigorous statements of incompressibility in (2.8)-(2.11) we will also use a slightly more restrictive but operationally sound definition that in an incompressible flow $\rho(\boldsymbol{x}, t) = \text{constant}$.

2.2.2 Advection-diffusion of a scalar

If a conserved scalar ϕ, in an isotropic, incompressible (constant density) continuum, has units of [conserved quantity/volume], and the flux vector can be expressed in terms of the diffusive flux

$$\boldsymbol{q} = -\kappa \nabla \phi \tag{2.12}$$

where the scalar diffusivity κ has units [area/time], the conservation (2.6) can be written as

$$\frac{d}{dt}\int_V \phi dV = \int_V Q \ dV + \int_A \kappa \nabla \phi \cdot n \, dA - \int_A (\boldsymbol{v} \cdot \boldsymbol{n}) \phi \, dA \qquad (2.13)$$

where Q [quanity/volume-time] is a volume source term. Taking the time derivative inside the volume integral (recall V is fixed in space) and using the divergence theorem we arrive at

$$\int_V \frac{\partial \phi}{\partial t} + \nabla \cdot (\boldsymbol{v}\phi) - \nabla \cdot (\kappa \nabla \phi) dV - Q = 0$$

Or since the volume V is chosen arbitrarily the argument has to be identically zero at every point in the domain, i.e.,

$$\frac{\partial \phi}{\partial t} + \nabla \cdot (\boldsymbol{v}\phi) - \nabla \cdot (\kappa \nabla \phi) - Q = 0 \qquad (2.14)$$

which should be immediately recognized as the point form of the well known transient advection diffusion equation. The derivation of (2.14) is made to emphasize the connection between the integral forms of the conservation equations, like those given in (2.6) and (2.13), and the more familiar and pervasive point forms like (2.14). In choosing between these two alternative balance statements, however, it will be repeatedly stressed throughout this work that the integral forms are the most suitable for developing numerical solutions based on Control Volume Finite Element technologies.

2.2.3 Stress and strain in an elastic solid

In determining the mechanical state (stresses and strains) in a body subjected to a loading, if we restrict attention to cases where translational and rotational rigid body motions do not occur and the body deforms but does not flow, we need only consider the linear momentum balance

$$\int_V \rho \boldsymbol{b} dV + \int_A \underline{\underline{\sigma}} \boldsymbol{n} dA = 0 \qquad (2.15)$$

obtained by setting the velocity $v = 0$ in (2.5). If we interpret the stress as the measure of change from an initial state of equilibrium and assume the body force remains unchanged during this evolution, the body force term can be dropped to arrive at

$$\int_A \underline{\underline{\sigma}} \boldsymbol{n} dA = 0 \tag{2.16}$$

Through the divergence theorem this equation can be written as a single volume integral, in index notation

$$\int_V \frac{\partial \sigma_{ij}}{\partial x_j} dV = 0 \tag{2.17}$$

or, since V is arbitrarily chosen, the point form

$$\frac{\partial \sigma_{ij}}{\partial x_j} = 0 \tag{2.18}$$

This last equation, the well known form of the equilibrium equation written in terms of change in stress, and is an often used starting point for the development of numerical solutions of solid mechanics problems. In this work, however, it is more convenient and direct to retain the alternative integral form in (2.16) or (2.17).

In further developing the statement (2.17) it is first noted that, if the loading of an isotropic, linear elastic body results in small deformations (compared to the dimension of the body as a whole) the stress can be related to the strain through the "generalized" Hooke's law;

$$\sigma_{ij} = \left[\frac{E\nu}{(1+\nu)(1-2\nu)} (\varepsilon_{kk} - \varepsilon_{kk}^0) \right] \delta_{ij} + \frac{E}{1+\nu} (\varepsilon_{ij} - \varepsilon_{ij}^0) \tag{2.19}$$

where, due to the small deformation assumption, the relationship between strains ε_{ij} and displacements u_i is

$$\varepsilon_{ij} = \frac{1}{2} \left(\frac{\partial u_j}{\partial x_i} + \frac{\partial u_i}{\partial x_j} \right) \tag{2.20}$$

and

$$\varepsilon_{kk} = \frac{\partial u_k}{\partial x_k} = \left(\frac{\partial u_x}{\partial x} + \frac{\partial u_y}{\partial y} + \frac{\partial u_z}{\partial z} \right) \tag{2.21}$$

In the above E is the Young's modulus [force/area], ν is Poisson's ratio, and $\delta_{ij}=1$ if $i=j$, $\delta_{ij}=0$ if $i \neq j$ is the Kronecker delta. The tensor ε_{ij}^0 is an initial strain—a strain independent of the stress; e.g., an isotropic thermal expansion,

$$\varepsilon_{ij}^0 = \alpha \Delta T \delta_{ij} \tag{2.22}$$

where α is the thermal expansion coefficient and ΔT is the temperature change from a reference state.

Substitution of (2.19) - (2.22) into (2.16) results in an integral equation in terms of the displacements. In component form, for a Cartesian coordinate system

$$\int_A \frac{E\nu}{(1+\nu)(1-2\nu)} \left[\frac{1-\nu}{\nu}\frac{\partial u_x}{\partial x} + \frac{\partial u_y}{\partial y} + \frac{\partial u_z}{\partial z} - \frac{1+\nu}{\nu}\alpha\Delta T \right] n_x$$
$$+ \frac{E}{2(1+\nu)} \left[\frac{\partial u_y}{\partial x} + \frac{\partial u_x}{\partial y} \right] n_y + \frac{E}{2(1+\nu)} \left[\frac{\partial u_z}{\partial x} + \frac{\partial u_x}{\partial z} \right] n_z \, dA = 0$$

$$\int_A \frac{E}{2(1+\nu)} \left[\frac{\partial u_y}{\partial x} + \frac{\partial u_x}{\partial y} \right] n_x$$
$$+ \frac{E\nu}{(1+\nu)(1-2\nu)} \left[\frac{\partial u_x}{\partial x} + \frac{1-\nu}{\nu}\frac{\partial u_y}{\partial y} + \frac{\partial u_z}{\partial z} - \frac{1+\nu}{\nu}\alpha\Delta T \right] n_y \tag{2.23}$$
$$+ \frac{E}{2(1+\nu)} \left[\frac{\partial u_z}{\partial y} + \frac{\partial u_y}{\partial z} \right] n_z \, dA = 0$$

$$\int_A \frac{E}{2(1+\nu)} \left[\frac{\partial u_z}{\partial x} + \frac{\partial u_x}{\partial z} \right] n_x + \frac{E}{2(1+\nu)} \left[\frac{\partial u_z}{\partial y} + \frac{\partial u_y}{\partial z} \right] n_y \, dA$$
$$+ \frac{E\nu}{(1+\nu)(1-2\nu)} \left[\frac{\partial u_x}{\partial x} + \frac{\partial u_y}{\partial y} + \frac{1-\nu}{\nu}\frac{\partial u_z}{\partial z} - \frac{1+\nu}{\nu}\alpha\Delta T \right] n_z \, dA = 0$$

Equation (2.23) is a general statement for the displacements in a three-dimensional body. Two important, two-dimensional cases of this equation are considered in this work.

2.2.4 Plane stress

If the domain of the problem is much thinner in a given direction, z-say, the stresses in that direction can be neglected, i.e.,

$$\sigma_{xz} = \sigma_{yz} = \sigma_{zz} = 0 \tag{2.24}$$

This condition, referred to as the plane stress condition, essentially reduces solution of (2.23) to a two-dimensional problem in the x-y plane. On using the stress-strain relationship in (2.19) and the strain definition in (2.20) and (2.21) the conditions in (2.24) lead to the following relationships in displacements

$$\frac{\partial u_z}{\partial x} + \frac{\partial u_x}{\partial z} = 0, \tag{2.25}$$

$$\frac{\partial u_z}{\partial y} + \frac{\partial u_y}{\partial z} = 0, \tag{2.26}$$

and

$$\frac{\partial u_z}{\partial z} = -\frac{\nu}{1-\nu}\frac{\partial u_x}{\partial x} - \frac{\nu}{1-\nu}\frac{\partial u_y}{\partial y} + \frac{1+\nu}{1-\nu}\alpha\Delta T \tag{2.27}$$

On substitution in (2.23) and recognizing, in the two-dimensional case, that the area integrals are over a closed curve, these relationships lead to a system in terms of displacements in the *x-y* plane alone, i.e.,

$$\oint_S \frac{E}{(1-\nu^2)}\left[\frac{\partial u_x}{\partial x} + \nu\frac{\partial u_y}{\partial y} - (1+\nu)\alpha\Delta T\right] n_x + \frac{E}{2(1+\nu)}\left[\frac{\partial u_y}{\partial x} + \frac{\partial u_x}{\partial y}\right] n_y \, dS = 0 \tag{2.28}$$

$$\oint_S \frac{E}{2(1+\nu)}\left[\frac{\partial u_y}{\partial x}+\frac{\partial u_x}{\partial y}\right]n_x + \frac{E}{(1-\nu^2)}\left[\nu\frac{\partial u_x}{\partial x}+\frac{\partial u_y}{\partial y}-(1+\nu)\alpha\Delta T\right]n_y dS = 0 \tag{2.29}$$

Or, following Fryer et al (1991), the more appropriate form for a control volume solution

$$\begin{aligned}&\oint_S \frac{E}{(1-\nu^2)}\frac{\partial u_x}{\partial x}n_x + \frac{E}{2(1+\nu)}\frac{\partial u_x}{\partial y}n_y\, dS = S_x\\&\oint_S \frac{E}{2(1+\nu)}\frac{\partial u_y}{\partial x}n_x + \frac{E}{(1-\nu^2)}\frac{\partial u_y}{\partial y}n_y\, dS = S_y\end{aligned} \tag{2.30}$$

Where

$$S_x = -\oint_S \frac{E}{(1-\nu^2)}\left[\nu\frac{\partial u_y}{\partial y}-(1+\nu)\alpha\Delta T\right]n_x + \frac{E}{2(1+\nu)}\frac{\partial u_y}{\partial x}n_y\, dS = 0$$

$$S_y = -\oint_S \frac{E}{2(1+\nu)}\frac{\partial u_x}{\partial y}n_x + \frac{E}{(1-\nu^2)}\left[\nu\frac{\partial u_x}{\partial x}-(1+\nu)\alpha\Delta T\right]n_y\, dS = 0$$

The in-plane stresses resulting from the displacement fields in (2.28) and (2.29) are

$$\begin{aligned}\sigma_{xx} &= \frac{E}{(1-\nu^2)}\left[\frac{\partial u_x}{\partial x}+\nu\frac{\partial u_y}{\partial y}-(1+\nu)\alpha\Delta T\right]\\ \sigma_{xy} &= \frac{E}{2(1+\nu)}\left[\frac{\partial u_y}{\partial x}+\frac{\partial u_x}{\partial y}\right]\\ \sigma_{yy} &= \frac{E}{(1-\nu^2)}\left[\nu\frac{\partial u_x}{\partial x}+\frac{\partial u_y}{\partial y}-(1+\nu)\alpha\Delta T\right]\end{aligned} \tag{2.31}$$

2.2.5 Plane strain

In contrast to the plane stress case, if the domain has a constant cross-section, much thicker in the *z*-direction, the strains (but not the stresses) can be neglected in that direction, i.e.,

$$\varepsilon_{xz} = \varepsilon_{yz} = \varepsilon_{zz} = 0 \tag{2.32}$$

This condition, referred to as the plane strain condition, can also reduce solution (2.23) to a two-dimensional problem in the x-y plane. Through the definition of strain in (2.20) and (2.21), the conditions (2.32) allows for (2.23) to be written in two closed equations for the displacements in the x-y plane, i.e.,

$$\oint_S \frac{E\nu}{(1+\nu)(1-2\nu)}\left[\frac{1-\nu}{\nu}\frac{\partial u_x}{\partial x} + \frac{\partial u_y}{\partial y} - \frac{1+\nu}{\nu}\alpha\Delta T\right]n_x + \frac{E}{2(1+\nu)}\left[\frac{\partial u_y}{\partial x} + \frac{\partial u_x}{\partial y}\right]n_y \, dS = 0 \tag{2.33a}$$

$$\oint_S \frac{E}{2(1+\nu)}\left[\frac{\partial u_y}{\partial x} + \frac{\partial u_x}{\partial y}\right]n_x + \frac{E\nu}{(1+\nu)(1-2\nu)}\left[\frac{\partial u_x}{\partial x} + \frac{1-\nu}{\nu}\frac{\partial u_y}{\partial y} - \frac{1+\nu}{\nu}\alpha\Delta T\right]n_y \, dS = 0 \tag{2.33b}$$

which can be rearranged

$$\oint_S \frac{E(1-\nu)}{(1+\nu)(1-2\nu)}\frac{\partial u_x}{\partial x}n_x + \frac{E}{2(1+\nu)}\frac{\partial u_x}{\partial y}n_y \, dS = S_x$$

$$\oint_S \frac{E}{2(1+\nu)}\frac{\partial u_y}{\partial x}n_x + \frac{E(1-\nu)}{(1+\nu)(1-2\nu)}\frac{\partial u_y}{\partial y}n_y \, dS = S_y \tag{2.34}$$

where, in this case,

$$S_x = -\oint_S \frac{E\nu}{(1+\nu)(1-2\nu)}\left[\frac{\partial u_y}{\partial y} - \frac{1+\nu}{\nu}\alpha\Delta T\right]n_x + \frac{E}{2(1+\nu)}\frac{\partial u_y}{\partial x}n_y \, dS = 0$$

$$S_y = -\oint_S \frac{E}{2(1+\nu)}\frac{\partial u_x}{\partial y}n_x + \frac{E\nu}{(1+\nu)(1-2\nu)}\left[\frac{\partial u_x}{\partial x} - \frac{1+\nu}{\nu}\alpha\Delta T\right]n_y \, dS = 0$$

The in-plane stresses resulting from the displacement fields in (2.33) are

$$\sigma_{xx} = \frac{E\nu}{(1+\nu)(1-2\nu)}\left[\frac{1-\nu}{\nu}\frac{\partial u_x}{\partial x} + \frac{\partial u_y}{\partial y} - \frac{1+\nu}{\nu}\alpha\Delta T\right]$$

$$\sigma_{xy} = \frac{E}{2(1+\nu)}\left[\frac{\partial u_y}{\partial x} + \frac{\partial u_x}{\partial y}\right] \tag{2.35}$$

$$\sigma_{yy} = \frac{E\nu}{(1+\nu)(1-2\nu)}\left[\frac{\partial u_x}{\partial x} + \frac{1-\nu}{\nu}\frac{\partial u_y}{\partial y} - \frac{1+\nu}{\nu}\alpha\Delta T\right]$$

2.2.6 Relationship between plane stress and plane strain

In developing codes to solve the plane stress and plane strain problems derived above it is not necessary to develop two separate codes. Rather, it is much more convenient to switch between plane stress and plane strain by simply defining "compound" material constants. For example, it is relatively easy to show (Barber (2003)) that the replacement of the Young's modulus, Possion's ratio, and thermal coefficient of expansion in the plane stress equations (2.29) by the compound constants

$$E^* = \frac{E}{1-\nu^2}, \quad \nu^* = \frac{\nu}{1-\nu}, \quad \alpha^* = (1+\nu)\alpha \tag{2.36}$$

results in the plane strain formulation in (2.33). Hence, if a plane stress code is available a given plane strain problem can be readily computed on using the compound material constants in (2.36) in place of the given constants.

2.2.7 The Navier-Stokes equations

In the previous treatment specific two-dimensional equations for the displacements in an elastic loaded body have been derived from a general three-dimensional form. The intention now is to repeat this exercise to arrive at a description of the flow field in a two-dimensional domain from a general three dimensional form.

To fully describe the velocity field in a flow we need to develop mass and momentum conservation equations. The basic case to study is the flow of an incompressible Newtonian flow (shear stress proportional to strain rate). The appropriate integral and point forms for the mass conservation are given in (2.8)-(2.11), hence the requirement here is to develop a momentum conservation equation in terms of the unknown velocity field, $\boldsymbol{v}$.

In general, the stress tensor $\underline{\underline{\sigma}}$ in a flowing fluid can be written as the sum of an isotropic part involving the pressure and a non-isotropic or deviatoric part involving the tangential stresses (Batchelor (1970)). In index notation

$$\sigma_{ij} = -p\delta_{ij} + 2\mu(\dot{\varepsilon}_{ij} - \tfrac{1}{3}\dot{\varepsilon}_{kk}\delta_{ij}) \tag{2.37}$$

where p is the pressure, and $\dot{\varepsilon}_{ij}$ is the strain-rate tensor—note the over-script dot used to distinguish this from the strain tensor ε_{ij} introduced in the linear elasticity equations— and μ is the dynamic viscosity. In a Newtonian fluid—the strain rate is related to the velocities

$$\dot{\varepsilon}_{ij} = \frac{1}{2}\left(\frac{\partial v_i}{\partial x_j} + \frac{\partial v_j}{\partial x_i}\right) \tag{2.38}$$

and, if the flow is incompressible (see (2.11)) the term

$$\dot{\varepsilon}_{kk} = \left(\frac{\partial v_k}{\partial x_k}\right) \equiv \frac{\partial v_x}{\partial x} + \frac{\partial v_y}{\partial y} + \frac{\partial v_w}{\partial z} = \nabla \cdot \boldsymbol{v} = 0 \tag{2.39}$$

In this way, on substituting for $\underline{\underline{\sigma}}$, the momentum balance (2.5) for a Newtonian incompressible fluid can be written as

$$\frac{d}{dt}\int_V \rho v_i dV = \int_V \rho b_i dV$$
$$-\int_A p n_i dA + \int_A \mu \frac{\partial v_i}{\partial x_j} n_j dA + \int_A \mu \frac{\partial v_j}{\partial x_i} n_j dA - \int_A \rho v_j n_j v_i dA \tag{2.40}$$

On using the divergence theorem, and, moving the time derivative under the integral sign (remember the volume V is fixed) (2.40) can be written solely in terms of an integration over the control volume

$$\int_V \frac{\partial(\rho v_i)}{\partial t} dV = \int_V \rho b_i dV$$
$$-\int_V \frac{\partial p}{\partial x_i} dV + \int_V \frac{\partial}{\partial x_j}\left(\mu \frac{\partial v_i}{\partial x_j}\right) dV + \int_V \frac{\partial}{\partial x_j}\left(\mu \frac{\partial v_j}{\partial x_i}\right) dV - \int_V \frac{\partial}{\partial x_j}(\rho v_i v_j) dV \tag{2.41}$$

If the viscosity is constant, this equation can be further simplified by switching the order of differentiation in the 4th term on the right hand side and noting that—due to the condition of incompressibility—$\partial v_j / \partial x_j \equiv \nabla \cdot \boldsymbol{v} = 0$ to arrive at

$$\int_V \frac{\partial(\rho v_i)}{\partial t} dV = \int_V \rho b_i dV$$
$$-\int_V \frac{\partial p}{\partial x_i} dV + \int_V \mu \frac{\partial}{\partial x_j}\left(\frac{\partial v_i}{\partial x_j}\right) dV - \int_V \frac{\partial}{\partial x_j}(\rho v_i v_j) dV \tag{2.42}$$

Since the volume V is arbitrary this can be written as

$$\frac{\partial(\rho v_i)}{\partial t} + \frac{\partial}{\partial x_j}(\rho v_i v_j) = \rho b_i - \frac{\partial p}{\partial x_i} + \mu \frac{\partial}{\partial x_j}\left(\frac{\partial v_i}{\partial x_j}\right) \tag{2.43}$$

which can be identified as the conserved point form of the Navier-Stokes equation for an incompressible constant viscosity flow. The non-conserved form, arrived at by expanding the left and side and using the incompressibility condition

$$\frac{\partial \rho}{\partial t} + \boldsymbol{v} \cdot \nabla \rho = 0,$$

is

$$\rho \frac{\partial v_i}{\partial t} + \rho v_j \frac{\partial v_i}{\partial x_j} = \rho b_i - \frac{\partial p}{\partial x_i} + \mu \frac{\partial}{\partial x_j}\left(\frac{\partial v_i}{\partial x_j}\right) \tag{2.44}$$

2.2.8 The stream-function—vorticity formulation

In this work we will only concern ourselves with solving steady, incompressible, constant property ($\rho, \mu = \text{constant}$) flow problems in two-dimensional (x-y) domains. For this case, in the absence of body forces, the momentum balance equation (2.42) can be written as

$$\int_A \nabla \cdot (v_x \boldsymbol{v}) + \frac{1}{\rho}\frac{\partial p}{\partial x} - \nu \nabla^2 v_x \, dA = 0 \tag{2.45a}$$

$$\int_A \nabla \cdot \left(v_y \boldsymbol{v}\right) + \frac{1}{\rho}\frac{\partial p}{\partial y} - \nu \nabla^2 v_y \, dA = 0 \tag{2.45b}$$

where $\nu = \mu / \rho$ is the kinematic viscosity. Note in this two-dimensional case the control volume V is now a control area A and the nabla symbol is defined as $\nabla \equiv (\partial / \partial x, \partial / \partial y)$. In seeking a solution of (2.45) it is convenient to eliminate the pressure p. This can be achieved by taking the partial derivative of (2.45a) with respect to y and subtracting the result from the partial derivate of (2.45b) with respect to x. After some manipulation and rearrangement this arrives at the single momentum conservation equation with the advection-diffusion form

$$\int_A \nabla \cdot (\boldsymbol{v}\omega) - \nu \nabla^2 \omega dA = 0 \tag{2.46}$$

where

$$\omega = \frac{\partial v_y}{\partial x} - \frac{\partial v_x}{\partial y} \tag{2.47}$$

is the vorticity. Progress is made by introducing the stream function defined by

$$v_x = \frac{\partial \Psi}{\partial y}, \; v_y = -\frac{\partial \Psi}{\partial x} \tag{2.48}$$

Note this definition automatically satisfies the incompressibility conditions (2.8)-(2.11) and allows the vorticity to be related to the stream function via (2.47)

$$\nabla^2 \Psi = -\omega \tag{2.49}$$

Appropriate integral forms suitable for a CVFEM solution can be obtained by using the divergence theorem to transform the area integrals into surface integrals. In particular (2.46) can be written as

$$\oint_S \omega \boldsymbol{v} \cdot \boldsymbol{n} - \nu \nabla \omega \cdot \boldsymbol{n} \, dS = 0 \tag{2.50}$$

and (2.49) can be written as

$$-\int_A \omega \, dA = \oint_S \nabla \psi \cdot \boldsymbol{n} \, dS \tag{2.51}$$

In a CVFEM application, an iterative solution, is constructed; for a given velocity field, a discrete form of (2.50) is solved for ω, followed by a solution of a discrete form of (2.51) for Ψ, and a velocity update by (2.48).

Chapter 3

The Essential Ingredients in a Numerical Solution

The essential ingredients in numerical solution for the field problems of Chapter 2 are outlined and discussed.

3.1 The Basic Idea

The field problems governed by the equations of Chapter 2 operate over a spatial domain with conditions, in terms of the dependent variables and their derivatives, prescribed on the boundary. A closed analytical solution of the governing equations provides a continuous function, such that for any given point in the problem domain values of the dependent variables can be calculated exactly. Closed form solutions, however, can only be derived for a limited number of special cases and, in general, a numerical solution needs to be constructed. The essential feature in a numerical solution is to obtain a discrete solution where values of the dependent variables are only obtained at a set of distinct points distributed throughout the domain—the node points; solution values at other points can be obtained by interpolation of the local node values. The nodal values of the dependent variables are obtained from the solution of an algebraic set of equations that relate the dependent variables at a given node point to the values at neighboring nodes. Appropriate algebraic equations—referred to as discrete equations—are obtained by one of two ways. The first approach uses suitable mathematical treatments of the governing equations in Chapter 2, e.g., finite difference approximations of terms in the equation, Smith (1985) or the method of weighted residuals, Zienkiewicz and Taylor (1989).

The second approach—and the one adopted in this work—is to arrive at discrete equations based on consideration of the same physical principals used to derive the continuous governing equations in Chapter 2. This approach is referred to as the Control Volume Method (CVM), Patankar (1980). The objective of this chapter is to provide a description of the key ingredients in developing a Control Volume solution.

3.2 The Discretization: Grid, Mesh, and Cloud

The initial step of arriving at the discrete equations is to place the node points into the domain. Broadly speaking there are three ways in which this can be achieved.

3.2.1 Grid

A basic approach assigns the location of nodes using a structured grid where, in a two-dimensional domain, the location of a node is uniquely specified by a row and a column index, see Fig. 3.1a. Although such a structured approach can lead to very convenient and efficient discrete equations it lacks flexibility in accommodating complex geometries or allowing for the local concentration of nodes in solution regions of particular interest.

3.2.2 Mesh

Geometric flexibility, usually at the expense of solution efficiency, can be added by using an unstructured mesh. Fig. 3.1b shows an unstructured mesh of triangular elements. In two dimensional domains triangular meshes are good choices because they can tessellate any planar surface. Note however, other choices of elements can be used in place of or in addition to triangular elements. The mesh can be used to determine the placement of the nodes. A common choice is to place the nodes at the vertices of the elements. In the case of triangles, this will allow for the approximation of a dependent variable, over the element, by linear

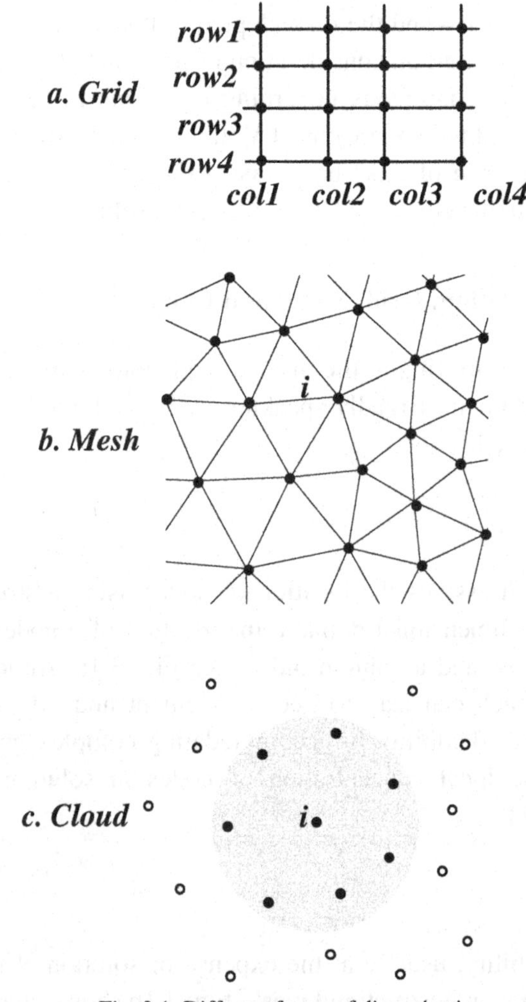

Fig. 3.1 Different forms of discretization

interpolation between the vertex nodes. Higher order approximations can be arrived at by using more nodes (e.g., placing nodes at mid points) and/or alternative elements (e.g., quadrilaterals).

In considering an unstructured mesh it is important to recognize that the following:

1. The quality of the numerical solution obtained is critically dependent on the mesh. For example, a key quality requirement for a mesh of

triangular elements is to avoid highly acute angles. The generation of appropriate meshes for a given domain is a complex topic worthy of a monograph in its own right. Fortunately, for two-dimensional problems in particular, there is a significant range of commercial and free software that can be used to generate quality meshes, e.g., the Delaunay triangularization routines in MATLAB used to create Fig. 3.1b.

2. The term unstructured is used to indicate a lack of a global structure that relates the position of all the nodes in the domain. In practice, however, a local structure—the region of support—listing the nodes connected to a given node i, is required. Establishing, storing and using this local data structure is one of the critical ingredients in using an unstructured mesh (see Chapter 4).

3.2.3 Cloud

The most flexible discretization is to simply populate the domain with node points that have no formal background grid or mesh connecting the nodes. Solution approaches based on this mesh-less form of discretization create local and structures, usually based on the "cloud" of neighboring nodes that fall with in a given length scale of a given node i, see Fig. 3.1c. A discussion of the features of meshless solutions can be found in Pepper (2006).

3.2.4 Discretizations for the Control Volume Finite Element Method

CVFEM solutions are based on a mesh of elements. The main discretization used in this work will be meshes of two-dimensional triangular elements (Fig. 3.1b). In the following discussion the components of this form of mesh and its role in a CVFEM solution are examined in detail.

3.3 The Element and the Interpolation Shape Functions

As noted previously, the building block of the discretization is the triangular element, Fig. 3.2. For linear triangular elements the node points are placed at the vertices. In Fig. 3.2 the nodes, moving in a counter-clockwise direction, are labeled 1, 2 and 3. Values of the dependent variable ϕ are calculated and stored at these node points. In this way, values at an arbitrary point (x, y) within the element can be approximated with linear interpolation

$$\phi \approx ax + by + c \tag{3.1}$$

where the constant coefficients *a*, *b* and *c* satisfy the nodal relationships

$$\phi_i = ax_i + by_i + c, \quad i = 1, 2, 3 \tag{3.2}$$

Equation (3.1) can be more conveniently written in terms of the shape function N_1, N_2, and N_3, where

$$N_i(x, y) = \begin{cases} 1, & \text{at node i} \\ 0, & \text{at all points on side opposite node i} \end{cases} \tag{3.3}$$

$$\sum_{i=1}^{3} N_i(x, y) = 1, \quad \text{at every point in the element} \tag{3.4}$$

such that, over the element, the continuous unknown field can be expressed as the linear combination of the values at nodes $i = 1, 2, 3$

$$\phi(x, y) \approx \sum_{i=1}^{3} N_i(x, y)\phi_i \tag{3.5}$$

With linear triangular elements a straightforward geometric derivation for the shape functions can be obtained. With reference to Fig. 3.2, observe that the area of the element is given by

$$A^{123} = \frac{1}{2}\begin{vmatrix} 1 & x_1 & y_1 \\ 1 & x_2 & y_2 \\ 1 & x_3 & y_3 \end{vmatrix}$$

$$= \frac{1}{2}[(x_2 y_3 - x_3 y_2) - x_1(y_3 - y_2) + y_1(x_3 - x_2)] \tag{3.6}$$

and the area of the sub-elements with vertices at points $(p, 2, 3)$, $(p, 3, 1)$, and $(p, 1, 2)$, where p is an arbitrary and variable point in the element, are given by

$$A^{p23} = \frac{1}{2}[(x_2 y_3 - x_3 y_2) - x_p(y_3 - y_2) + y_p(x_3 - x_2)] \quad (3.7a)$$

$$A^{p31} = \frac{1}{2}[(x_3 y_1 - x_1 y_3) - x_p(y_1 - y_3) + y_p(x_1 - x_3)] \quad (3.7b)$$

$$A^{p12} = \frac{1}{2}[(x_1 y_2 - x_2 y_1) - x_p(y_2 - y_1) + y_p(x_2 - x_1)] \quad (3.7c)$$

With these definitions it follows that the shape functions are given by

$$N_1 = A^{p23} / A^{123}, N_2 = A^{p31} / A^{123}, N_3 = A^{p12} / A^{123} \quad (3.8)$$

Note that, when point p coincides with node i (=1, 2 or 3) the shape function $N_i = 1$, and when point p is anywhere on the element side opposite node i, the associated sub-element area is zero, and through (3.8), the shape function $N_i = 0$. Hence the shape functions defined by (3.8) satisfy the required condition in (3.3). Further, note that at any point p, the sum of the areas

$$A^{p23} + A^{p31} + A^{p12} = A^{123}$$

such that the shape functions at (x_p, y_p) will sum to unity. Hence the shape functions defined by (3.8) also satisfies the condition (3.4).

For future reference it is worthwhile to note that the derivatives of the shape functions in (3.8) over the element are the following constants

$$\begin{aligned}
N_{1x} &= \frac{\partial N_1}{\partial x} = \frac{(y_2 - y_3)}{2A^{123}}, \quad N_{1y} = \frac{\partial N_1}{\partial y} = \frac{(x_3 - x_2)}{2A^{123}} \\
N_{2x} &= \frac{\partial N_2}{\partial x} = \frac{(y_3 - y_1)}{2A^{123}}, \quad N_{2y} = \frac{\partial N_2}{\partial y} = \frac{(x_1 - x_3)}{2A^{123}} \\
N_{3x} &= \frac{\partial N_3}{\partial x} = \frac{(y_1 - y_2)}{2A^{123}}, \quad N_{3y} = \frac{\partial N_3}{\partial y} = \frac{(x_2 - x_1)}{2A^{123}}
\end{aligned} \quad (3.9)$$

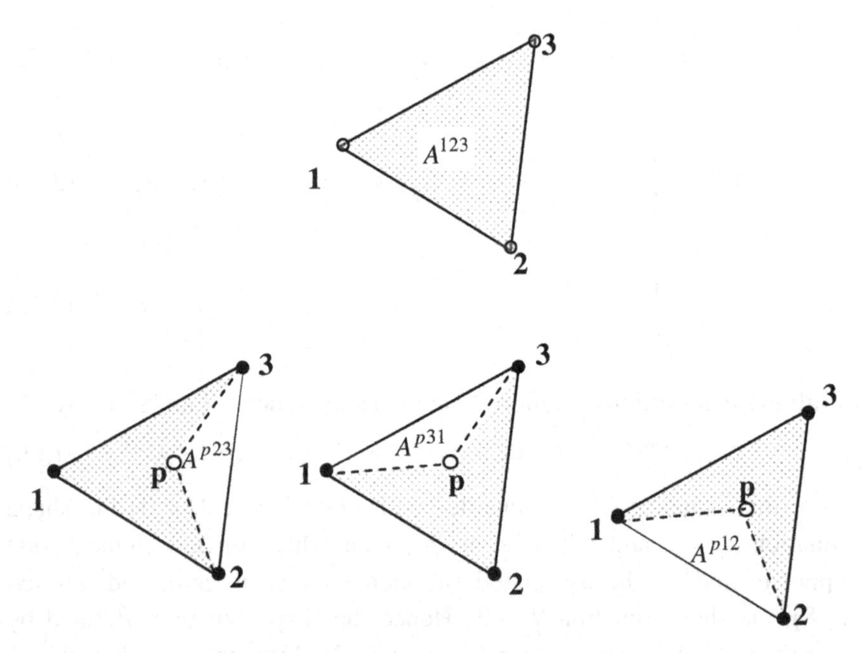

Fig. 3.2 An element indicating the areas used in shape function definitions

3.4 Region of Support and Control Volume

The local structure on the mesh in Fig. 3.1b is defined in terms of the region of support—the list of nodes that share a common element with a given node i, in Fig. 3.3. In this region of support, as illustrated in Fig. 3.3, a control volume is created by joining the center of each element in the support to the mid points of the element sides that pass through node i. This creates a closed polygonal control volume with $2m$ sides (faces); where m is the number of elements in the support. Each element contributes 1/3 of its area to the control volume area and the volumes from all the nodes tessellate the domain without over lap. As noted below, in arriving at the discrete numerical equation, the control volume is related to the domain of integration in the integral forms of the governing equations in Chapter 2.

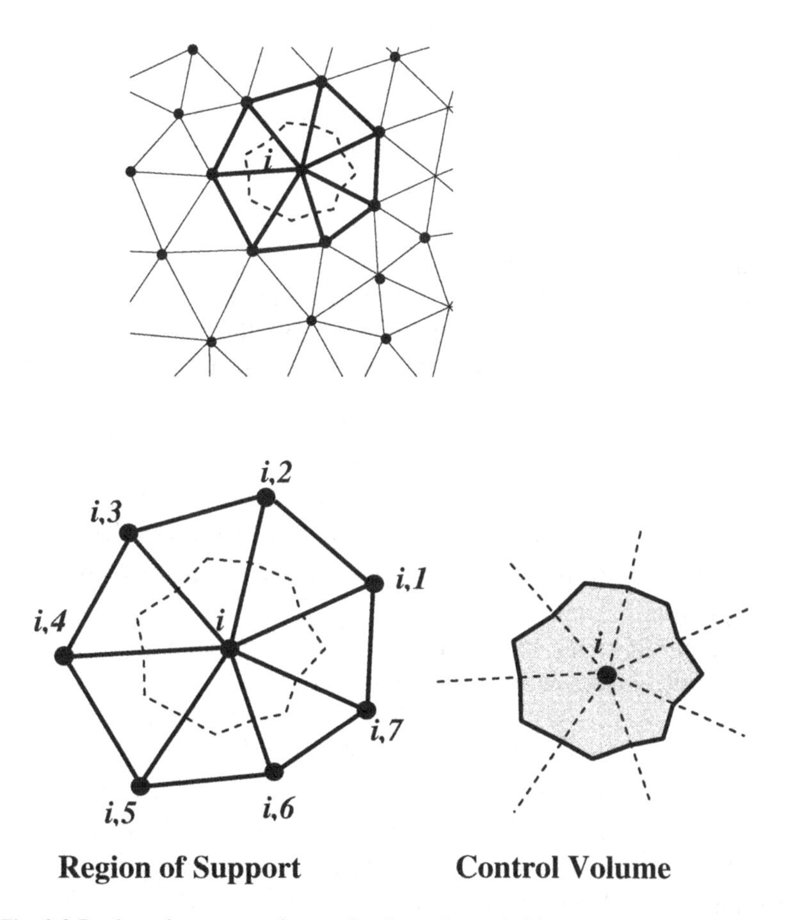

Fig. 3.3 Region of support and control volume for node i in an unstructured mesh of linear triangular elements

3.5 The Discrete Equation

With the definitions provided above we are now in position to outline the main steps in reducing the continuous description of the governing equations in Chapter 2 into an algebraic discrete form. As a guiding example consider a transient diffusion problem. In integral form the governing equation, adapted from (2.12) assuming a two dimensional domain Ω, is

$$\frac{d}{dt}\int_A \phi dA = \oint_S \kappa \nabla \phi \cdot n\, dS, \quad x \in \Omega \tag{3.10}$$

As indicated in Fig. 3.4a, the domain in the integral (3.10) can be any arbitrary closed area in Ω, including areas that share common surfaces with the boundary Γ of Ω.

The steps in converting (3.10) into a set of discrete algebraic equations in terms of the nodes distributed thought out Ω are as follows.

1. The domain is covered by a mesh of triangular elements, Fig. 3.4b.
2. The regions of support and control volumes attached to each node i are identified; this requires that an appropriate data structure is used, see Chapter 4.
3. The domains of the integrals in (3.10) are associated with the regions of support, Fig. 3.4b.
4. Using numerical integration (usually one-point rules) and the shape function approximations (3.5) of ϕ in each element of the ith support, equation (3.10) is expanded in terms of the nodal values of ϕ in the support.
5. On gathering terms, the resulting equation for node i can be written in the general discrete form

$$a_i \phi_i = \sum_{nb} a_{nb} \phi_{nb} + b_i \tag{3.11}$$

where a_i and a_{nb} are coefficients for the unknown nodal values of ϕ, and the additional coefficient b_i accounts for contributions from sources, transients and boundaries. Equation (3.11) provides an algebraic relationship between the value of ϕ at node i and the neighboring (nb) nodes in its support.

The central task of this text is to provide a detailed "recipe" for the determination of the coefficients in (3.11) for a given governing field equation and domain mesh. The main approach used is based on the control volume concept laid out above. Note, however, that any numerical method based on a domain discretization (grid, mesh, or cloud) can, regardless of the mathematical sophistication employed, be written in the form (3.11).

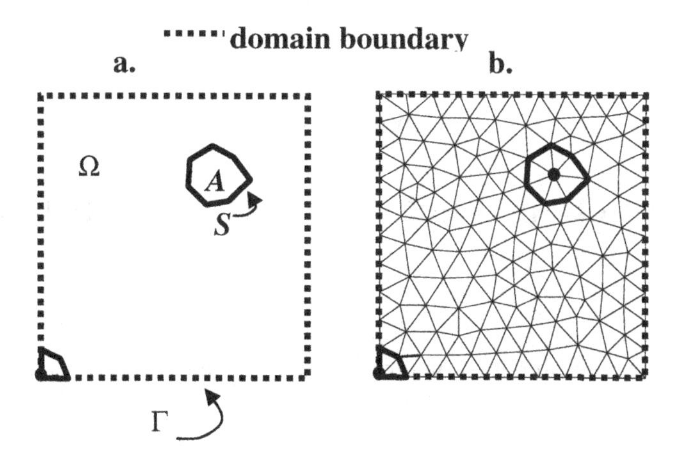

Fig. 3.4 Association between area domain in integral forms of the governing equation and the components of the discretization mesh

Chapter 4

Control Volume Finite Element Data Structure

An unstructured mesh data structure for control volume finite elements is presented.

4.1 The Task

In Chapter 3 the essential components in a domain discretization, i.e., nodes, mesh, elements, region of support, and control volume, were defined. This chapter outlines how these components are used to reduce a continuous governing field equation into a set of discrete equations in terms of unknown nodal values. In order to automate this process through the use of a computer code it is necessary to provide a data structure that can identify and store the mesh components (elements, support, control volume) associated with a given node point in the solution domain. Providing a suitable data structure for a Control Volume Finite Element Method (CVFEM) code is the current objective. Emphasis is placed on the term suitable to indicate that there is not a unique form for the data structure. In this respect, it is noted that the data structure presented here is not designed for efficiency or compactness but for clarity of presentation.

4.2 The Mesh

The starting point for the data structure presented here is to cover the solution domain with a mesh of linear triangles. As noted in Chapter 3, there are a number of commercial and open source codes that will generate high quality unstructured meshes that will be compatible with the data structure used in this work. The basic requirements are that these mesh generators provide

1. a contiguous and unique numbering (labeling) of the nodes ($i = 1,2,...,n$), where n is the number of nodes in the domain discretization,
2. a vector of the nodal x and y positions, i.e., x_i and y_i,
3. a contiguous unique numbering of the triangles in the discretization.
4. A listing of the node numbers (labels) of the vertices of the triangles (preferably in counter-clockwise order), and
5. a listing of the nodes that lie on the boundary of the domain.

It is recognized that it may not be possible or easy for a reader to find an unstructured mesh generator. To compensate for this a meshing code, based on a structured grid, is provided in Appendix A. This code, discussed in more detail in Appendix A, provides a "ready-to-use" mesh conforming to the required data structure.

4.3 The Data Structure

There are two central components in the data structure, the region of support for a node i in the domain discretization and the domain boundaries.

4.3.1 The region of support

Figure 4.1 shows a domain mesh which possesses the five informational attributes noted in the previous section. For each node i in the mesh the region of support is identified by counting and listing all the neighboring nodes k that share a common element side with node i. The number of neighboring nodes in the region of support is stored in the variable n_i^s where the index $i = 1,2...,n$. The nodes in the region of support of node i are stored in the two-dimensional array $S_{i,j}$, where the index $i = 1,2...,n$ and the index $j = 1,2,...n_i^s, n_i^s + 1$. For a specific example refer to Fig. 4.1

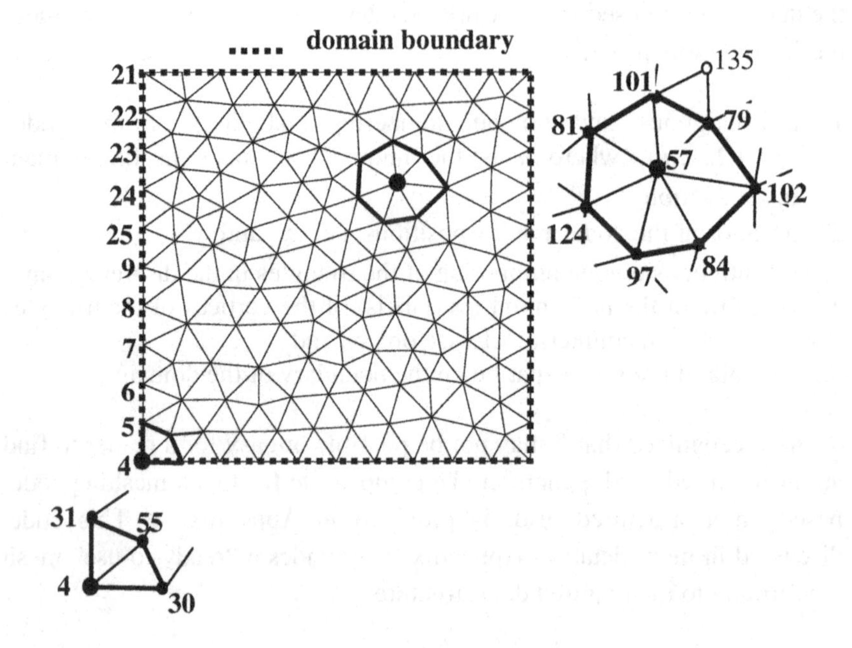

Fig. 4.1 Indices associated with regions of support in domain mesh

For an internal node, e.g., node $i = 57$

$$
\begin{aligned}
n^s_{57} = 7,\quad S_{57,1} = 102,\ S_{57,2} = 79,\ S_{57,3} = 101,\ S_{57,4} = 81,\\
S_{57,5} = 124,\ S_{57,6} = 97,\ S_{57,7} = 84,\ S_{57,8} = 102
\end{aligned}
\tag{4.1}
$$

For a boundary node, e.g., $i = 4$

$$
n^s_4 = 3,\quad S_{4,1} = 30,\ S_{4,2} = 55,\ S_{4,3} = 31,\ S_{4,4} = 0 \tag{4.2}
$$

Note also the following aspects of the region of support:

1. The support nodes are listed by moving counter-clockwise around the node i.
2. For internal nodes the first node in the support can be chosen arbitrarily.
3. For boundary nodes the list of support nodes must start and end on a boundary and the path must remain in the domain.
4. The additional storage location S_{i,n^s_i+1} is used to distinguish between internal and boundary nodes. At an internal node $S_{i,n^s_i+1} = S_{i,1}$

indicating complete enclosure of node i. At a boundary node $S_{i,n_i^e+1} = 0$.

4.3.2 The boundary

In addition to the region of support it is also important to store information on the domain boundaries. First the boundary is segmented into n^{seg} contiguous sections, identified by a common boundary condition, e.g., an imposed value or an imposed flux. Then, for a given boundary segment the global node numbers on the boundary are stored in the two-dimensional array $B_{k,j}$ where the index k is over the boundary segments, i.e., $k = 1, 2, ... n^{seg}$ and the index j is over the nodes on the boundary, i.e., $j = 1, 2, ... n_{B,k}$ ($n_{B.k}$ = the number of nodes on boundary segment).

Assume the 4 obvious boundary segments in the domain in Fig 4.1 and number these segments counter clockwise starting from the lower horizontal boundary. In this case the data structure for the left hand vertical boundary is

$$
\begin{aligned}
&n_4^b = 11, \\
&B_{4,1} = 21, B_{4,2} = 22, B_{4,3} = 23, B_{4,4} = 24, \\
&B_{4,5} = 25, B_{4,6} = 9, \ B_{4,7} = 8, \ B_{4,8} = 7, \\
&B_{4,9} = 6, \ B_{4,10} = 5, B_{4,11} = 4
\end{aligned}
\tag{4.3}
$$

Again note the counter clockwise counting of the boundary segments and the nodes on the segment.

4.4 The Discrete Equation

The key objective of a CVFEM is to reduce the integral from of a governing field equation, e.g.,

$$\frac{d}{dt}\int_A \phi dA = \oint_S \kappa \nabla \phi \cdot n \, dS, \quad \boldsymbol{x} \in \Omega \tag{4.4}$$

to a set of discrete algebraic equations in the unknown nodal values of ϕ. With the data structure given above the form of this equation, derived from (3.11) is

$$a_i\phi_i = \sum_{j=1}^{N_i} a_{i,j}\phi_{S_{i,j}} + b_i \tag{4.5}$$

where a_i is the coefficient associated with the unknown at node i, $a_{i,j}$ is the coefficient associated with the jth node in the ith support, and b_i accounts for contributions from sources, transients and boundaries. Figure 4.2 summarizes how the data structure and geometric components of the numerical mesh are used to automate the process of extracting (4.4) from (4.3). The steps in this automation are as follows.

1. For a given element in the support of i, a local node numbering [1, 2, 3] is employed. In the jth element of support i this local numbering is related to the global nodes through the identity
$$[1, 2, 3] = [i, S_{i,j}, S_{i,j+1}] \tag{4.6}$$
Note that local node 1 is always associated with node i and the remaining two nodes in the element are numbers in a counter clockwise fashion.
2. For each node j in the support of node i, the element in (4.6) is identified. Since the global node numbers of the vertices are known through (4.6), the local coordinates of the vertices in the element can be accessed by pointing to
$$\begin{aligned} x_1 &= x_{i,} \quad x_2 = x_{S_{i,j}}, \quad x_3 = x_{S_{i,j+1}} \\ y_1 &= y_i, \quad y_2 = y_{S_{i,j}}, \quad y_3 = y_{S_{i,j+1}} \end{aligned} \tag{4.7}$$
3. Equation (4.6) allows for the calculation of all the geometric features of the element. In particular, (i) its area A^j, (ii) the contribution of this area ($\frac{1}{3}A^j$) to the control volume area A_i,(iii) its linear shape functions N_1, N_2, N_3 and their derivatives, (iv) the unit normal on the faces of the control volume that reside in the element, and (v) the lengths of these faces. As detailed in Chapter 5, this is sufficient information to carry out a numerical integration of the surface

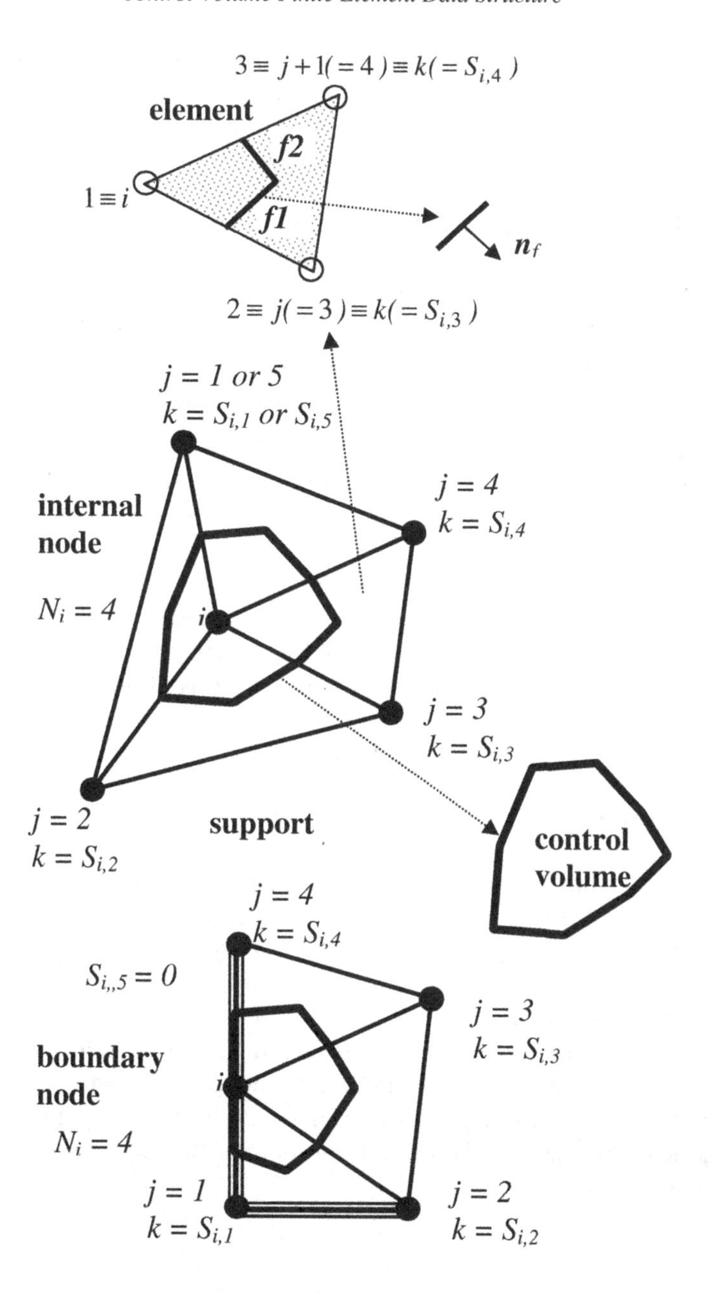

Fig. 4.2 Data structure and geometric components of the numerical mesh

integral on the right hand side of (4.4). A step that essentially approximates the fluxes crossing the two control volume faces, $f1$ and $f2$ in the element in question in terms of the nodal values $\phi_i, \phi_{S_{i,j}}, \phi_{S_{i,j+1}}$.

4. When the loop around the support of node is complete, i.e., the above calculations have been carried out for each element in the domain, the right hand side of (4.4) is fully determined. That is, the net flux crossing the faces of the control volume around node i are fully determined in terms of the nodal values at node i and its support. At the same time, summing the contributions ($\frac{1}{3}A^j$) from each element the area of the control volume is known. This value can be immediately used in a mid-point integration of the left hand side of (4.4) that will add additional dependencies on the nodal values to (4.5).
5. Appropriate gathering of the terms from steps 3 and 4 will allow for the identification of the coefficients in the general discrete equation (4.4). The process can be repeated for each node in the domain to arrive at closed systems of equations in terms of all the nodal values of ϕ.

4.5 Summary

This chapter has proposed a data structure for the use with a CVFEM code and discussed how this data structure can be used to automatically generate the CVFEM discrete equations in (4.5). The five steps identified in the previous section, however, only provide a sketch of how the proposed data structure is used to obtain the coefficients in (4.5). Many important details are not fully considered in this sketch, e.g., how to handle boundaries, source, transients, advection transport, variable properties, etc. Exposing these details and fully testing the resulting numerical schemes will be the central topic of the remaining chapters in this work.

Chapter 5

Control Volume Finite Element Method (CVFEM) Discretization and Solution

The key steps in obtaining and solving the discrete CVFEM equations, related to the solution of advective-diffusive problems are outlined.

5.1 The Approach

The object here is to detail the CVFEM discretization and solution procedures for the two-dimensional (x, y) advection-diffusion equation. In integral form the governing equation for the scalar $\phi(x, y)$ is

$$\frac{d}{dt}\int_V \phi dV - \int_V Q \; dV - \int_A \kappa \nabla \phi \cdot \boldsymbol{n} \, dA + \int_A (\boldsymbol{v} \cdot \boldsymbol{n}) \phi \, dA = 0 \tag{5.1}$$

In arriving at the appropriate discretization a step-by-step process is employed. Where possible, the development will closely follow how we might construct and write a code. Hence, although the text may be tedious and long-winded, it provide a reliable template for rapid code development.

Recall that the aim of the CVFEM discretization is to transform the description in (5.1) into a system of algebraic equations in terms of unknown values located at the node points (vertices) of a linear triangular finite element mesh. As discussed in detail in Chapters 3 and 4, the algebraic equation for node i in this mesh, relating the value at node i to it neighbors $(j = 1..n_i)$, has the form

$$a_i \phi_i = \sum_{j=1}^{n_i} a_{i,j} \phi_{S_{i,j}} + b_i \tag{5.2}$$

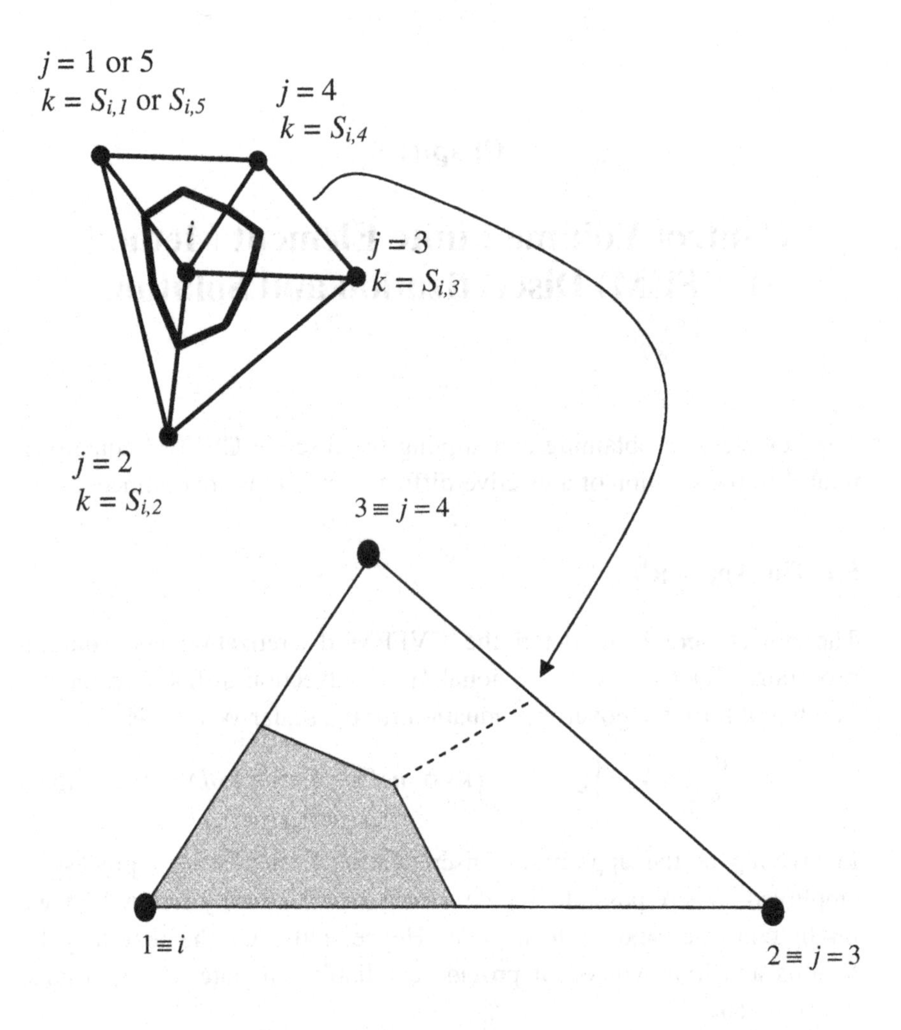

Fig. 5.1 Region of support for node i and the jth element in the support

where, with reference to Fig. 5.1, n_i is the number of nodes in the support (neighborhood)of node i , the a's are coefficients, $a_{i,j}$ is the coefficient of the jth node in the support, the index value $S_{i,j}$ gives the global nodal number (k) of the jth node in the support, and b_i accounts for contributions from sources, boundaries and transients. Critical aspects of constructing the coefficients in (5.2) can be fully illustrated by considering a single finite element in the support, see the lower part of Fig. 5.1. The coefficient calculations in this given element will be identical to the calculations used in all the other elements in the support. Then, on aggregating the calculated coefficient and source contributions from each element in turn, full specification of the coefficient and source terms in (5.2) can be achieved.

5.2 Preliminary Calculations

Before the coefficients associated with the selected single element in Fig. 5.1 are determined some preliminary values need to be calculated and stored.

The volume of the element (assuming unit depth) is given by

$$V^{ele} = \frac{1}{2}\begin{vmatrix} 1 & x_1 & y_1 \\ 1 & x_2 & y_2 \\ 1 & x_3 & y_3 \end{vmatrix}$$

$$= \frac{(x_2 y_3 - x_3 y_2) + x_1(y_2 - y_3) + y_1(x_3 - x_2)}{2} \tag{5.3}$$

The contribution of the volume of this element (the 3rd element in the region of support) to the control volume associated with node i (shown as the bold polygon in the upper image in Fig. 5.1 is

$$V_i^j = \tfrac{1}{3} V^{ele} \tag{5.4}$$

From Chapter 3 (Eq. (3.9)) the derivatives of the shape functions (constant in the elements) are

$$\begin{aligned}
N_{1x} &= \frac{\partial N_1}{\partial x} = \frac{(y_2 - y_3)}{2V^{ele}}, \quad N_{1y} = \frac{\partial N_1}{\partial y} = \frac{(x_3 - x_2)}{2V^{ele}} \\
N_{2x} &= \frac{\partial N_2}{\partial x} = \frac{(y_3 - y_1)}{2V^{ele}}, \quad N_{2y} = \frac{\partial N_2}{\partial y} = \frac{(x_1 - x_3)}{2V^{ele}} \\
N_{3x} &= \frac{\partial N_3}{\partial x} = \frac{(y_1 - y_2)}{2V^{ele}}, \quad N_{3y} = \frac{\partial N_3}{\partial y} = \frac{(x_2 - x_1)}{2V^{ele}}
\end{aligned} \tag{5.5}$$

From (5.5) the components of the gradient

$$\nabla\phi = \left(\frac{\partial\phi}{\partial x}, \frac{\partial\phi}{\partial y}\right)$$

in the element are

$$\frac{\partial\phi}{\partial x} = N_{1x}\phi_1 + N_{2x}\phi_2 + N_{3x}\phi_3 \tag{5.6a}$$

and

$$\frac{\partial\phi}{\partial y} = N_{1y}\phi_1 + N_{2y}\phi_2 + N_{3y}\phi_3 \tag{5.6b}$$

With reference to Fig. 5.2 and moving in a counter-clockwise direction around node i the change in x value, along face 1 in the element is

$$\Delta\vec{x}_{f1} = \frac{x_3}{3} - \frac{x_2}{6} - \frac{x_1}{6} \tag{5.7}$$

The change in y value is

$$\Delta\vec{y}_{f1} = \frac{y_3}{3} - \frac{y_2}{6} - \frac{y_1}{6} \tag{5.8}$$

The equivalent changes in x and y values along face 2 are respectively

$$\Delta\vec{x}_{f2} = -\frac{x_2}{3} + \frac{x_3}{6} + \frac{x_1}{6} \tag{5.9}$$

$$\Delta\vec{y}_{f2} = -\frac{y_2}{3} + \frac{y_3}{6} + \frac{y_1}{6} \tag{5.10}$$

The area of face 1 in the element (assuming unit depth) is

$$A_{f1} = \sqrt{\Delta\vec{x}_{f1}^2 + \Delta\vec{y}_{f1}^2} \tag{5.11}$$

The area of face 2 is

$$A_{f2} = \sqrt{\Delta\vec{x}_{f2}^2 + \Delta\vec{y}_{f2}^2} \tag{5.12}$$

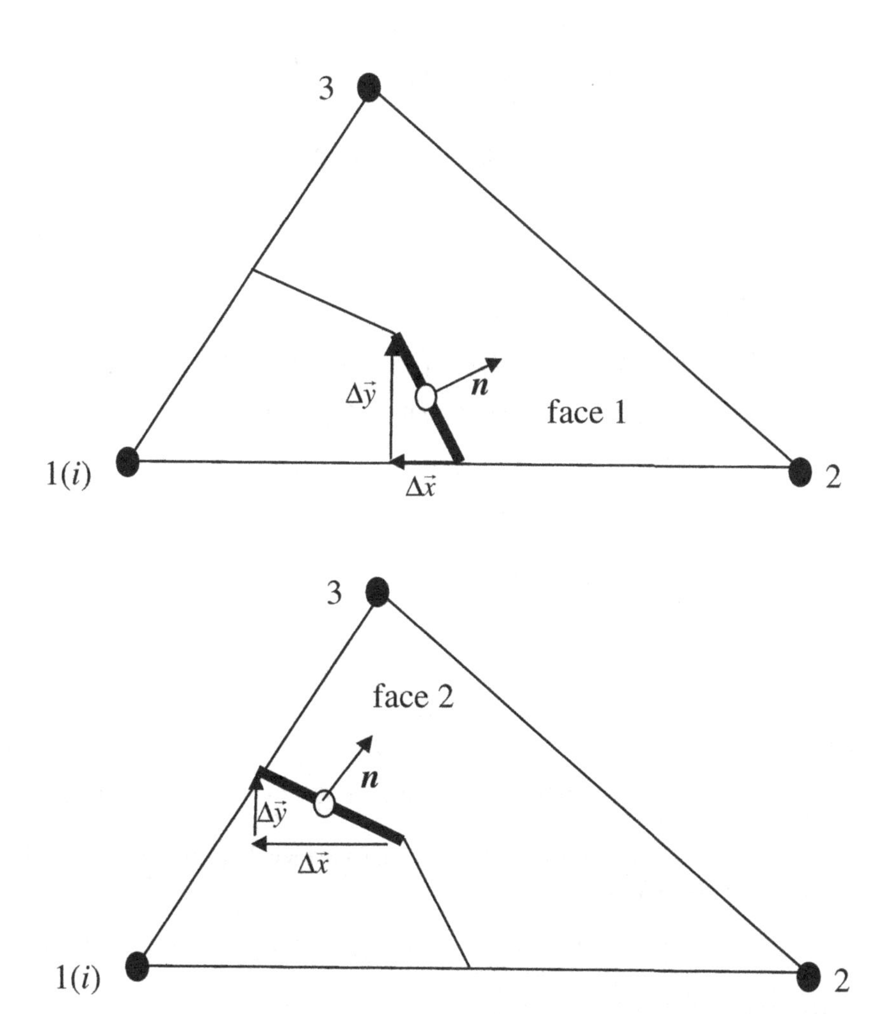

Fig. 5.2 The faces of the control volume

The components of the unit normal (pointing out of the ith control volume) on face 1 (see Fig. 5.2) is

$$n_x^{f1} = \frac{\Delta\vec{y}_{f1}}{A_{f1}}, \quad n_y^{f1} = \frac{-\Delta\vec{x}_{f1}}{A_{f1}} \tag{5.13}$$

On face 2 the components are

$$n_x^{f2} = \frac{\Delta\vec{y}_{f2}}{A_{f2}}, \quad n_y^{f2} = \frac{-\Delta\vec{x}_{f2}}{A_{f2}} \tag{5.14}$$

The value of the diffusivity at the mid-point of face 1 (indicated in Fig. 5.2) is

$$\kappa_{f1} = [N_1\kappa_1 + N_2\kappa_2 + N_3\kappa_3]_{f1} = \tfrac{5}{12}\kappa_1 + \tfrac{5}{12}\kappa_2 + \tfrac{2}{12}\kappa_3 \tag{5.15}$$

At the mid-point of face 2

$$\kappa_{f2} = [N_1\kappa_1 + N_2\kappa_2 + N_3\kappa_3]_{f1} = \tfrac{5}{12}\kappa_1 + \tfrac{2}{12}\kappa_2 + \tfrac{5}{12}\kappa_3 \tag{5.16}$$

The velocity components at the midpoint of face 1 are

$$v_x^{f1} = \tfrac{5}{12}v_{x_1} + \tfrac{5}{12}v_{x_2} + \tfrac{2}{12}v_{x_3} \tag{5.17a}$$

and

$$v_y^{f1} = \tfrac{5}{12}v_{y_1} + \tfrac{5}{12}v_{y_2} + \tfrac{2}{12}v_{y_3} \tag{5.17b}$$

On face 2

$$v_x^{f2} = \tfrac{5}{12}v_{x_1} + \tfrac{2}{12}v_{x_2} + \tfrac{5}{12}v_{x_3} \tag{5.18a}$$

and

$$v_y^{f2} = \tfrac{5}{12}v_{y_1} + \tfrac{2}{12}v_{y_2} + \tfrac{5}{12}v_{y_3} \tag{5.18b}$$

The volume flow (volume/time) *out* across face 1, in the direction of the outward normal, is

$$q_{f1} = \boldsymbol{v}\cdot\boldsymbol{n}A\big|_{f1} = v_x^{f1}\Delta\vec{y}_{f1} - v_y^{f1}\Delta\vec{x}_{f1} \tag{5.19}$$

Across face 2

$$q_{f2} = \boldsymbol{v}\cdot\boldsymbol{n}A\big|_{f2} = v_x^{f2}\Delta\vec{y}_{f2} - v_y^{f2}\Delta\vec{x}_{f2} \tag{5.20}$$

Armed with these preliminary calculations the appropriate coefficients in 5.2 can be calculated. This process is illustrated by developing discrete forms of the advection-diffusion equation.

5.3 Steady State Diffusion

For steady state diffusion the governing equation is

$$\int_A \kappa \nabla \phi \cdot \boldsymbol{n}\, dA = 0 \tag{5.21}$$

Associating the domain of this integration with the area of the control volume associated with node i (the bolded polygon in. Fig 5.1)

$$\int_A \kappa \nabla \phi \cdot \boldsymbol{n}\, dA = \sum_{j=1}^{n_i} \int_{A^j} \kappa \nabla \phi \cdot \boldsymbol{n}\, dA = 0 \tag{5.22}$$

where A^j is the area of the control volume faces in the jth element of the support. Note that at an internal domain node the number of elements and nodes in a support are equal. Equation (5.22) states that the net diffusive flux into the ith control volume is zero.

Each term in the sum on the right-hand side of (5.22) can be approximated using a midpoint integration rule. For the selected triangular element in Fig. 5.1 this approximation leads to

$$\begin{aligned} \int_{A^3} \kappa \nabla \phi \cdot \boldsymbol{n}\, dA &= \int_{face1} \kappa \nabla \phi \cdot \boldsymbol{n}\, dA + \int_{face2} \kappa \nabla \phi \cdot \boldsymbol{n}\, dA \\ &\approx \kappa \nabla \phi \cdot \boldsymbol{n}\, A\big|_{f1} + \kappa \nabla \phi \cdot \boldsymbol{n}\, A\big|_{f2} \end{aligned} \tag{5.23}$$

From the results in the previous section the following expansions can be made

$$\begin{aligned} \kappa \nabla \phi \cdot \boldsymbol{n}\, A\big|_{f1} &= \kappa_{f1} \frac{\partial \phi}{\partial x} \Delta \vec{y}_{f1} - \kappa_{f1} \frac{\partial \phi}{\partial y} \Delta \vec{x}_{f1} \\ &= \kappa_{f1} [\, N_{1x}\phi_1 + N_{2x}\phi_2 + N_{3x}\phi_3 \,] \Delta \vec{y}_{f1} \\ &\quad - \kappa_{f1} [\, N_{1y}\phi_1 + N_{2y}\phi_2 + N_{3y}\phi_3 \,] \Delta \vec{x}_{f1} \end{aligned} \tag{5.24}$$

and

$$\begin{aligned} \kappa \nabla \phi \cdot \boldsymbol{n}\, A\big|_{f2} &= \kappa_{f2} \frac{\partial \phi}{\partial x} \Delta \vec{y}_{f2} - \kappa_{f2} \frac{\partial \phi}{\partial y} \Delta \vec{x}_{f2} \\ &= \kappa_{f2} [\, N_{1x}\phi_1 + N_{2x}\phi_2 + N_{3x}\phi_3 \,] \Delta \vec{y}_{f2} \\ &\quad - \kappa_{f2} [\, N_{1y}\phi_1 + N_{2y}\phi_2 + N_{3y}\phi_3 \,] \Delta \vec{x}_{f2} \end{aligned} \tag{5.25}$$

Substituting these expansions in (5.23) and using global indexing for the unknown nodal field results in

$$\int_{A^3} \kappa \nabla \phi \cdot \boldsymbol{n}\, dA \approx -a_1^{\kappa}\phi_i + a_2^{\kappa}\phi_{S_{i,3}} + a_3^{\kappa}\phi_{S_{i,4}} \tag{5.26}$$

where

$$\begin{aligned}
a_1^{\kappa} &= -\kappa_{f1}N_{1x}\Delta\vec{y}_{f1} + \kappa_{f1}N_{1y}\Delta\vec{x}_{f1} - \kappa_{f2}N_{1x}\Delta\vec{y}_{f2} + \kappa_{f2}N_{1y}\Delta\vec{x}_{f2} \\
a_2^{\kappa} &= \kappa_{f1}N_{2x}\Delta\vec{y}_{f1} - \kappa_{f1}N_{2y}\Delta\vec{x}_{f1} + \kappa_{f2}N_{2x}\Delta\vec{y}_{f2} - \kappa_{f2}N_{2y}\Delta\vec{x}_{f2} \\
a_3^{\kappa} &= \kappa_{f1}N_{3x}\Delta\vec{y}_{f1} - \kappa_{f1}N_{3y}\Delta\vec{x}_{f1} + \kappa_{f2}N_{3x}\Delta\vec{y}_{f2} - \kappa_{f2}N_{3y}\Delta\vec{x}_{f2}
\end{aligned} \tag{5.27}$$

The superscript $()^{\kappa}$ is used to indicate that these values are the diffusive contribution to the coefficients in the general form (5.2). These values can be used to *update* the ith support coefficients through

$$\begin{aligned}
a_i &= a_i + a_1^{\kappa} \\
a_{i,3} &= a_{i,3} + a_2^{\kappa} \\
a_{i,4} &= a_{i,4} + a_3^{\kappa}
\end{aligned} \tag{5.28}$$

The *updating* form of (5.28) is used to indicate that contributions to a given support coefficient can come from other elements in the support; take care with this equation to recognize the distinction between indices identifying support nodes and nodes in the selected triangular element, e.g., $()_{i,4}$ represents the 4th node in the support of node i (upper part of Fig 5.1 and $()_3$ represents the 3rd node in the selected element (lower part of Fig. 5.1).

5.4 Steady State Advection-Diffusion

For Steady State advection-diffusion the governing equation is

$$\int_A \kappa \nabla \phi \cdot \boldsymbol{n}\, dA - \int_A (\boldsymbol{v} \cdot \boldsymbol{n})\phi\, dA = 0 \tag{5.29}$$

Associating the domain of this integration with the area of the control volume for node i (the bolded polygon in Fig. 5.1) allows for the expansion

$$\int_A \kappa\nabla\phi\cdot n\,dA-\int_A (v\cdot \boldsymbol{n})\phi\,dA$$

$$=\sum_{j=1}^{n_i}\int_{A^j}\kappa\nabla\phi\cdot n\,dA-\int_{A^j}(v\cdot \boldsymbol{n})\phi\,dA=0 \qquad (5.30)$$

This equation states that the net advective + diffusive flux into the ith control volume is zero. Each term in the sum on the right-hand side of (5.30) can be approximated using a midpoint integration rule. For the selected triangular element in Fig. 5.1 this approximation leads to

$$\int_{A^3}\kappa\nabla\phi\cdot n\,dA-\int_{A^3}(v\cdot \boldsymbol{n})\phi\,dA$$

$$=\int_{A^3}\kappa\nabla\phi\cdot \boldsymbol{n}\,dA-\int_{face1}(v\cdot \boldsymbol{n})\phi\,dA-\int_{face2}(v\cdot \boldsymbol{n})\phi\,dA \qquad (5.31)$$

$$\approx\int_{A^3}\kappa\nabla\phi\cdot \boldsymbol{n}\,dA-(v\cdot \boldsymbol{n})\phi A\big|_{f1}-(v\cdot \boldsymbol{n})\phi A\big|_{f2}$$

On expansion, the first term on the right-hand-side will generate the diffusion coefficients given in (5.27), and the later terms can be written in terms of the volume flow rates (see (5.19) and (5.20)) to arrive at

$$\int_{A^3}\kappa\nabla\phi\cdot \boldsymbol{n}\,dA-(v\cdot \boldsymbol{n})\phi A\big|_{f1}-(v\cdot \boldsymbol{n})\phi A\big|_{f2}$$

$$=-a_1^{\kappa}\phi_i+a_2^{\kappa}\phi_{S_{i,3}}+a_3^{\kappa}\phi_{S_{i,4}}-q_{f1}\phi_{f1}-q_{f2}\phi_{f2} \qquad (5.32)$$

where the a's are the diffusive coefficients given in (5.27). To move forward we need to be able to approximate the face values of the unknown scalar ϕ. An obvious choice is

$$\phi_{f1}=[N_1\phi_1+N_2\phi_2+N_3\phi_3]_{f1}=\tfrac{5}{12}\phi_1+\tfrac{5}{12}\phi_2+\tfrac{2}{12}\phi_3 \qquad (5.33a)$$

and

$$\phi_{f2}=[N_1\phi_1+N_2\phi_2+N_3\phi_3]_{f2}=\tfrac{5}{12}\phi_1+\tfrac{2}{12}\phi_2+\tfrac{5}{12}\phi_3 \qquad (5.33b)$$

This is equivalent to the central difference approximation for the advective face value in the Control Volume Finite Difference Method. In a similar manner to the FDCVM this choice can lead to a poorly conditioned system in advection dominated problems (Patankar (1980)).

The first order avoidance measure for this problem is to employ an upwind approach, setting

$$\phi_{f1} = \begin{cases} \phi_1 & if \quad q_{f1} > 0 \\ \phi_2 & if \quad q_{f1} < 0 \end{cases} \tag{5.34a}$$

and

$$\phi_{f2} = \begin{cases} \phi_1 & if \quad q_{f2} > 0 \\ \phi_3 & if \quad q_{f2} < 0 \end{cases} \tag{5.34b}$$

This will always lead to well conditioned systems but can also induce numerical dissipation (artificial diffusion). Similar concepts to those used in CVFDM can be used to improve CVFEM performance for advection dominated flows but due to the unstructured nature of the mesh they can be a little trickier to implement, e.g., see Woodfield et al (2004).

If upwinding is used the last two terms in (5.32) can be replaced with

$$-q_{f1}\phi_{f1} = -\max[q_{f1},0]\phi_1 + \max[-q_{f1},0]\phi_2 \tag{5.35a}$$

and

$$-q_{f2}\phi_{f2} = -\max[q_{f2},0]\phi_1 + \max[-q_{f2},0]\phi_2 \tag{5.35b}$$

On making this replacement and gathering terms, we can write down the following approximation for the advective + diffusive flux arriving into the control volume across the faces located in element 3

$$\int_{A^3} \kappa\nabla\phi \cdot \boldsymbol{n}\, dA - (\boldsymbol{v}\cdot\boldsymbol{n})\phi A\big|_{f1} - (\boldsymbol{v}\cdot\boldsymbol{n})\phi A\big|_{f2}$$

$$= -(a_1^{\kappa} + a_1^{u})\phi_i + (a_2^{\kappa} + a_2^{u})\phi_{S_{i,3}} + (a_3^{\kappa} + a_3^{u})\phi_{S_{i,4}} \tag{5.36}$$

where the advective coefficients, identified with the superscripts $()^u$, are given by

$$\begin{aligned} a_1^u &= \max[q_{f1},0] + \max[q_{f2},0] \\ a_2^u &= \max[-q_{f1},0] \\ a_3^u &= \max[-q_{f2},0] \end{aligned} \tag{5.37}$$

These values can be used to *update* the ith support coefficients through

$$
\begin{aligned}
a_i &= a_i + a_1^\kappa + a_1^u \\
a_{i,3} &= a_{i,3} + a_2^\kappa + a_2^u \\
a_{i,4} &= a_{i,4} + a_3^\kappa + a_3^u
\end{aligned}
\tag{5.38}
$$

The updating form is used to account for contributions to the nodal coefficients from other elements in the support. Again, as noted below equation (5.28), take care to recognize the distinction between indices identifying support nodes and nodes in the selected triangular element.

5.5 Steady State Advection-Diffusion with Source Terms

When source terms are present the governing equation is

$$
\int_V Q \; dV - \int_A \kappa \nabla \phi \cdot \boldsymbol{n} \, dA + \int_A (\boldsymbol{v} \cdot \boldsymbol{n}) \phi \, dA = 0 \tag{5.39}
$$

The discretization of the diffusive and advection terms (terms 2 and 3 on the left hand side) are detailed in the two previous sections. Here we just focus on the source term. If the volume V is associated with the control volume around node i in Fig. 5.1 this term can be written

$$
\int_{V_i} Q \; dV = \sum_{j=1}^{\text{elements}} \int_{V_i^j} Q \, dV \tag{5.40}
$$

where the sum is over the elements in the region of support for node i and V_i^j is the volume contribution to the control volume from the jth element in the support, see (5.4).

5.5.1 Volume source terms

If $Q(x,y)$ is a volume source term (quantity/volume-time) then the right hand side can be approximated by a one-point integration to arrive at

$$
\sum_{j=1}^{\text{elements}} \int_{V_i^j} Q \, dV \approx Q_i V_i \tag{5.41}
$$

where $Q_i = Q(x_i, y_i)$ is the value of the source evaluated at the location of node i and

$$V_i = \sum_{j=1}^{elements} V_i^j$$

is the volume of the ith control volume. This approach of approximating the source is referred to as nodal lumping and is effectively a one-point integral centered on the node point of the given control volume. An alternative is to perform separate volume integrals for each element on the right-hand-side of (5.40). The domain for such an integral is the shaded region illustrated in the lower part of Fig 5.1 and can be performed using a Gauss quadrature. This approach, referred to as a consistent approach, may have more accuracy if $Q(x,y)$ changes rapidly. Nodal lumping, however, results in better convergence properties for the system (5.2) and is often the preferred approach in CVFEM

5.5.2 Source linearization

When $Q(x,y)$ is a volume source term the full discrete equation corresponding to (5.39) is

$$a_i \phi_i = \sum_{j=1}^{n_i} a_{i,j} \phi_{S_{i,j}} + Q_i V_i \tag{5.42}$$

It is possible that, in the general case, $Q_i = Q(x_i, y_i)$ will be a function of the nodal unknown value ϕ_i. Thus (5.42) could form a non-linear system and its solution could require iteration. A crude approach simply calculates $Q_i = Q(x_i, y_i)$ based on the best available estimates for ϕ_i $(i = 1.. n)$ and then solve the system (5.42) to arrive at a better estimates for ϕ_i. If the source is highly non-linear, however, convergence of this approach could be very slow or non-existent. Computational solution performance can be dramatically improved by linearizing the source term

$$Q_i V_i = -Q_{C_i} \phi_i + Q_{B_i} \tag{5.43}$$

In (5.43) Q_C and Q_B can still be functions of the nodal values of ϕ but to be effective it is *essential* that $Q_C > 0$ throughout the calculation; if (5.42) is written as the system $\boldsymbol{A}\boldsymbol{\varphi} = \boldsymbol{b}$ a positive value $Q_C > 0$ will increase the magnitude of the diagonal elements of $\boldsymbol{A}$ and improve the conditioning of the system.

5.5.3 Line source

If the source is a line source located at $\boldsymbol{x}_k = (x_k, y_k)$,

$$Q = Q_\ell \delta(\boldsymbol{x}_k - \boldsymbol{x}) \tag{5.44}$$

where Q_ℓ (quantity/time-unit depth) is independent of ϕ, and $\delta(\boldsymbol{x}_k - \boldsymbol{x})$, the Dirac delta function, has the properties

$$\delta(\boldsymbol{x}_k - \boldsymbol{x}) = \begin{cases} 0, & \boldsymbol{x} \neq \boldsymbol{x}_k \\ \infty, & \boldsymbol{x} = \boldsymbol{x}_k \end{cases} \tag{5.45a}$$

$$\int_{-\infty}^{\infty} \delta(\boldsymbol{x}_k - \boldsymbol{x}) = 1 \tag{5.45b}$$

In the case that an internal node i is located on $\boldsymbol{x}_k$, i.e., $\boldsymbol{x}_i = \boldsymbol{x}_k$, see Fig. 5.3a

$$\int_{V_i} Q \; dV = Q_\ell \tag{5.46}$$

and the terms in the linearized source are

$$Q_{C_i} = 0, \;\text{ and }\; Q_{B_i} = Q_l \tag{5.47}$$

If the node i is located on an insulated (symmetry) boundary, however, see Fig 5.3b, the source term is

$$\int_{V_i} Q \; dV = \frac{\theta}{2\pi} Q_\ell \tag{5.48}$$

where θ is the angle (radians) between the boundaries joining at node i; this results in the settings

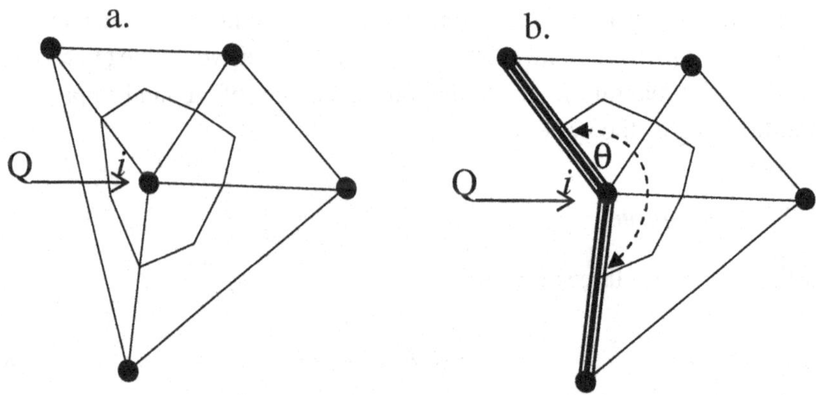

Fig. 5.3 Treatment of line source; a. node i located inside domain, b. node i located on insulated boundary

$$Q_{C_i} = 0, \quad \text{and } Q_{B_i} = \frac{\theta}{2\pi} Q_l \tag{5.49}$$

5.6 Coding Issues

The discrete form of the steady-state advection-diffusion equation (5.39) is

$$[\, a_i + Q_{C_i} \,]\phi_i = \sum_{j=1}^{n_i} a_{i,j}\phi_{S_{i,j}} + Q_{B_i} \tag{5.50}$$

The work in prior parts of this chapter have shown how the terms in this equation can be calculated using CVFEM. At this point is worthwhile to summarize these calculations by outlining how a computational code to evaluate the terms in (5.50) can be constructed.

Using the data structure and nomenclature outlined in Chapter 4.

An outer loop calls at each of the n nodes in the mesh, $i = 1..n$.

-- At each node i values for the support volume V_i, and the coefficients Q_{C_i}, Q_{B_i} and a_i, $a_{i,j}$ ($j = 1..n_i$) are set to a default value of zero.

-- Each node i is identified as an internal node point ($S_{i,n_i+1} = S_{i,1} \neq 0$) or a boundary node point ($S_{i,n_i+1} = 0$) and a loop counter is set to $loop = n_i$ at internal nodes or $loop = n_i - 1$ at boundary nodes.

-- A loop is made over the nodes in the support of node i, $j = 1 .. loop$.

[**Note:** the value of j in this loop can be associated with the jth element in the support of node i. Internal node supports will always have the same number of elements as support nodes, hence the setting $loop = n_i$. Boundary node supports will always have one less element than support node, hence the setting $loop = n_i - 1$]

----At each value of j in the loop:

----The coordinates of the vertices of the jth element are extracted

$$x_1 = x_i, \quad x_2 = x_{S_{i,j}}, \quad x_3 = x_{S_{i,j+1}}$$

$$y_1 = y_i, \quad y_2 = y_{S_{i,j}}, \quad y_3 = y_{S_{i,j+1}}$$

----The preliminary data is calculated, (5.3) to (5.20), and stored

----The control volume is updates as $V_i = V_i + V_i^j$

----The diffusive coefficients a_1^k, a_2^k, a_3^k are calculated (5.27) and stored

----The advective coefficients a_1^u, a_2^u, a_3^u are calculated (5.37) and stored

---- The support coefficients are updated from

$$a_i = a_i + a_1^{\kappa} + a_1^u$$

$$a_{i,j} = a_{i,j} + a_2^{\kappa} + a_2^u$$

$$\begin{cases} a_{i,j+1} = a_{i,j} + a_3^{\kappa} + a_3^u & \text{if} \quad j \neq n_i \\ a_{i,1} = a_{i,j} + a_3^{\kappa} + a_3^u & \text{if} \quad j = n_i \end{cases}$$

[**Note:** the last term is constructed to account for the fact that in an internal node the 3rd node of the last element (element $j = n_i$) is the 1st node in the support.]

--The loop on the nodes in the support closes. This loop has completely calculated the coefficients in for the ith equation in the system (5.50).

--Volume source contributions to (5.50) are calculated with (5.41) using if required the linearization in (5.43)

--Line source contributions are calculated with (5.47) or (5.49).

The loop on the nodes in the domain closes. At this point (5.50) is fully established for each node $i = 1 .. n$ in the domain.

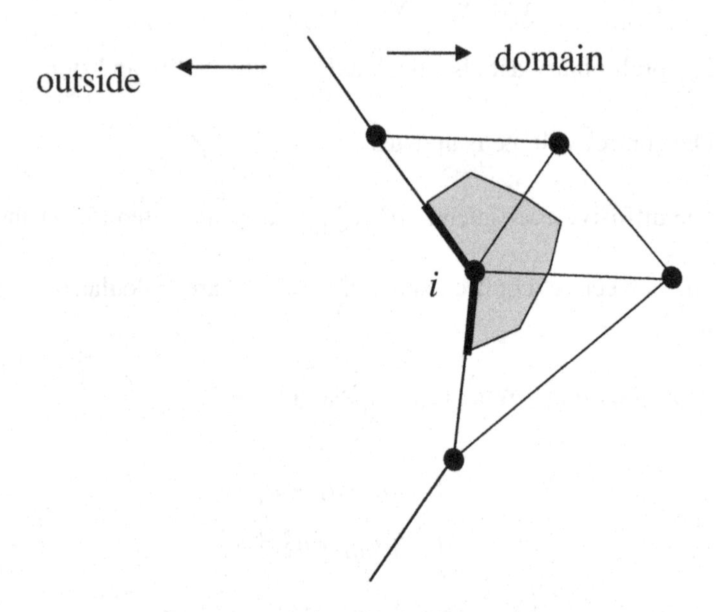

Fig. 5.4 A typical domain boundary node point

5.7 Boundary Conditions

Up to this point, the terms in the discrete form for a steady state advection-diffusion equation have been fully defined. Before a real problem can be solved, however, appropriate boundary conditions need to be applied. This step requires the modification of (5.50) with the addition of extra terms for the nodes i that are located on a domain boundary. Consider a typical boundary node, Fig 5.4, located on the jth node $(1 \le j \le n_{B,k})$ of the $(1 \le k \le n^{seg})$ domain boundary segment. Recall from the data structure given in Chapter 4, that n^{seg} is the number of segments of the domain boundary (arranged counterclockwise around the domain boundary), $n_{B,k}$ is the number of nodes on the kth segment, and the counter j sequentially counts the nodes in the counterclockwise direction. The discrete equation associated with the node i in Fig. 5.4 is a balance over the shaded polygonal control volume. Key parts of this balance are the fluxes crossing the control volume faces. In previous sections the diffusive and convective fluxes across the faces that do not belong to the domain boundaries have been evaluated. Contributions across control faces that coincide with the domain boundary—the bold lines in Fig. 5.4—have, however, not been determined.

5.7.1 Face area calculations

To provide a general treatment for boundary conditions some preliminary calculation of the boundary face areas associated with each node j in a given boundary segment are required. Figure 5.5 shows a schematic of the kth $(k = 3)$ boundary, indicating the data structure. Assuming unit depth, the face area associated with any node j of the boundary segment highlighted in Fig 5.5 is given by

$$A_{k,j} = \begin{cases} Upper_1 \\ Upper_j + Lower_j, \quad 2 \le j \le n_{B.k} - 1 \\ Lower_{n_{B,k}} \end{cases} \tag{5.51}$$

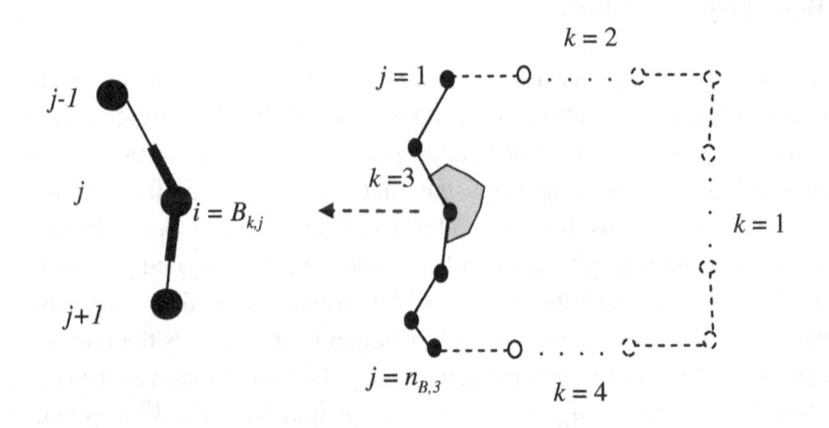

Fig. 5.5 A domain boundary segment with three sequential points detailed

where

$$Upper_j = \tfrac{1}{2}\sqrt{(x_j - x_{j+1})^2 + (y_j - y_{j+1})^2},$$
$$Lower_j = \tfrac{1}{2}\sqrt{(x_j - x_{j-1})^2 + (y_j - y_{j-1})^2}$$

The first and last lines on the right-hand side of (5.51) account for the first and last node on the boundary segment. This treatment assumes that there are at least two boundary segments—i.e., at least two contiguous regions of the domain boundary where different boundary conditions are applied. In cases where only one boundary condition is applied (e.g., a constant value over the whole boundary) the condition of a two-segment boundary can be artificially imposed.

The coding to calculate the boundary face values can be as follows:

Loop over the boundary segments $k = 1 .. n_k$

--Loop over the nodes on segment k, $j = 1 .. n_{B,k}$

----Associate local boundary coordinates with global coordinates, i.e.,

$$x_{j-1} = x_{B_{k,j-1}},\ y_{j-1} = y_{B_{k,j-1}},\quad j > 1$$
$$x_j = x_{B_{k,j}},\ y_j = y_{B_{k,j}},\qquad 1 \le j < n_{B,k}$$
$$x_{j+1} = x_{B_{k,j+1}},\ y_{j+1} = y_{B_{k,j+1}},\quad j < n_{B,k}$$

----Calculate $A_{k,j}$ from (5.51)

--Close loop on j

Close loop on k

5.7.2 Convective condition

A general treatment for boundary conditions can be achieved by considering the convective boundary condition

$$\int_A q_{in}\, dA = \int_A h(\,\phi_{amb} - \phi\,)\, dA \tag{5.52}$$

This states that the net flux (quantity/area-time) entering the domain across the area *A* is specified by a transfer coefficient $h \ge 0$ and an ambient value of the field variable ϕ_{amb}. When the domain of the integral, *A*, is associated with the face areas of node *i*, coinciding with the domain boundary, this is the flux value that needs to be added to the right-hand side of (5.50). Using a nodal lumping approach the integral on the right of (5.52) can be approximated as

$$\int_A h(\,\phi_{amb} - \phi\,)\, dA \approx -B_{C_i}\phi_i + B_{B_i} \tag{5.53}$$

where

$$B_{C_i} = hA_{k,j} > 0, \quad B_{B_i} = h\phi_{amb}A_{k,j} \tag{5.54}$$

and the index $i \equiv B_{k,j}$. In this way the convective boundary condition can be added by extending (5.50) to

$$[\,a_i + Q_{C_i} + B_{C_i}\,]\phi_i = \sum_{j=1}^{n_i} a_{i,j}\phi_{S_{i,j}} + Q_{B_i} + B_{B_i} \tag{5.55}$$

This can be achieved by the following coding:

Loop over the nodes in the domain, $i = 1..n$, setting $B_{C_i} = 0$ and $B_{B_i} = 0$

Loop over the boundary segments $k = 1..n_k$

--Loop over the nodes on segment k, $j = 1..n_{B,k}$

---- Calculate global node index associated with node *j*, i.e., $i = B_{k,j}$

----Calculate B_{C_i} and B_{B_i} from (5.54)

--Close loop over *j* segment

Close loop on boundary segments *k*

5.7.3 Generalization of the convective boundary condition

By appropriate choice of the value of h and ϕ_{amb} the definition of the boundary source terms, B_{C_i} and B_{B_i}, can be effectively used to model a range of possible boundary conditions.

Convective Boundary: Clearly a finite setting of h and ϕ_{amb} will provide, by definition, a convective boundary condition.

Insulated Boundary: A setting of $h = 0$ and a finite value of ϕ_{amb} will provides an insulated –no-flow—boundary. The only coding required for this boundary condition is to ensure that, for nodes i on an insulated boundary, the settings $B_{C_i} = 0$ and $B_{B_i} = 0$ are made. Note this is the default condition arising from CVFEM procedures leading to the discrete from (5.50).

Fixed Value Boundary: A setting of $h = 10^{16}$ (a large number) and $\phi_{amb} = \phi_{value}$ (a prescribed value) ensures that at the boundary node i in question, the terms B_{C_i} and $B_{B_i} = B_{C_i}\phi_{value}$ will dominate, essentially reducing (5.55) to

$$B_{C_i}\phi_i \approx B_{B_i} \tag{5.56}$$

On noting that $B_{B_i} = B_{C_i}\phi_{value}$, this choice forces the fixed value (Dirichlet) condition $\phi_i \approx \phi_{value}$. In practice this prescribed-value boundary condition can be obtained with a minimum of coding. All that it required is that, for nodes i on a fixed value boundary, the settings $B_{C_i} = 10^{16}$ and $B_{B_i} = \phi_{value} \times 10^{16}$ are made.

Fixed Flux Boundary: Finally a setting of $h = 10^{-16}$ (a small value) and $\phi_{amb} = q_{in} / h$, where q_{in} is a prescribed flux into the domain, will essentially force the settings $B_{C_i} \approx 0$ and $B_{B_i} = A_{k,j}q_{in}$, such that, at global node i corresponding to node j on the prescribed flux boundary segment k, (5.55) takes the required form

$$[\, a_i + Q_{C_i} \,]\phi_i = \sum_{j=1}^{n_i} a_{i,j}\phi_{S_{i,j}} + Q_{B_i} + A_{k,j}q_{in} \tag{5.57}$$

5.8 Solution

The following sequential steps leads to a fully defined system of discrete equations, (5.55), for a given steady-state advection-diffusion problem:

-Calculation of diffusive coefficients
-Calculation of advective coefficients
-Treatment of sources
-Treatment of boundary conditions

The final step is to solve this equation (5.55) and determine the nodal field values $\phi_i, (i = 1..n)$. There is a plethora of possible solution techniques that could be applied. A flexible, if not necessarily efficient, approach is a point wise Gauss-Seidel iteration which writes (5.55) as

$$[\, a_i + Q_{C_i} + B_{C_i} \,]\phi_i^r = \left[\sum_{j=1}^{n_i} a_{i,j}\phi_{S_{i,j}} + Q_{B_i} + B_{B_i} \right]^{r,r-1} \tag{5.58}$$

where the superscript $()^r$ indicates the rth approximation and the subscript $[]^{r,r-1}$ indicates that the item in brackets is evaluated with the currently available approximation. Starting from an initial guess for the nodal field—$\phi_i^0 = 0, (i = 1..n)$ is often a reasonable default—(5.58) is solved to provide an updated values $\phi_i^1, (i = 1..n)$. In a subsequent step, these value can be used on the right hand side of (5.58) to obtain the updates $\phi_i^2, (i = 1..n)$. This iteration cycle continues until convergence, defined by

$$max\left|\phi_i^r - \phi_i^{r-1}\right|, (i = 1..n) < \text{specified tolerance} \qquad (5.59)$$

The equation in (5.58) could be non-linear—i.e., the coefficients (*a*), source terms (*Q*) and boundary conditions (*B*) could be functions of the unknown ϕ_i. In this case under-relaxation may be required to solve (5.58). The point solution is rewritten as

$$\phi_i^r = \phi_i^{r-1} + \omega\left(\frac{\left[\sum_{j=1}^{n_i} a_{i,j}\phi_{S_{i,j}} + Q_{B_i} + B_{B_i}\right]^{r,r-1} - [\, a_i + Q_{C_i} + B_{C_i}\,]\phi_i^{r-1}}{[\, a_i + Q_{C_i} + B_{C_i}\,]}\right)$$

(5.60)

where $0 < \omega \leq 1$ is a relaxation factor. Note that the numerator in the brackets on the right hand side determines, for the current field estimates, the miss-match between the left and the right hand side of (5.55). Hence the term in brackets can be viewed as a correction, improving the estimate from ϕ_i^{r-1} to ϕ_i^r; note, on convergence this correction vanishes. The convergence of (5.60) is controlled by the choice of ω, if too large a value is chosen the point iteration may diverge, if too small convergence may be very slow (require multiple iterations). The optimum value is usually found by trial and error. In this respect it is worth noting that if the system (5.55) is linear (or close to linear) the optimum value may be $\omega \geq 1$, referred to as over-relaxation.

Appropriate coding for the solver is as follows—throughout this code the best available value of the field variable $\phi_i^{r-1,r}$ is stored as ϕ_i and the updated value ϕ_i^r is stored as ϕ_i^{up}.
An initial guess is assigned to the field value $\phi_i = 0, (i = 1..n)$

Values for a convergence tolerance *tol*, an under-relaxation factor ω, and maximum number of iterations *maxit* are prescribed. A default for the maximum difference is set as *max-diff* = 0

A loop is made over an iteration counter $r = 1..\ maxit$

--For each choice of *r*, a loop is made over the domain nodes $i = 1..n$

----Looping over the nodes, $j = 1..n_i$, in the ith region of support

------The value $\sum_{j=1}^{n_i} a_{i,j}\phi_{S_{i,j}}$ of is calculated and stored.

----Close Loop on *j*

----Update the nodal value of $\phi_i^{up} = \phi_i^r$ from (5.60)

----If $\max\left|\phi_i^{up} - \phi_i\right| >$ *max-diff* set *max-diff* $= max\left|\phi_i^{up} - \phi_i\right|$

----Set current best nodal value $\phi_i = \phi_i^{up}$

--Close Loop on *i*

--If *max-diff*<*tol* terminate iterations

Close Loop on *r*

If *r* = *maxit*, print a warning that convergence has not been achieved

5.9 Handling Variable Diffusivity

In the treatment provided above, it is assumed that the diffusivity is a continuous and known function of space $\kappa = f(x, y)$. In this way nodal values κ_i can be calculated *a-priori* and values at any point in an element can be found through the shape function interpolation (5.15) and (5.16). There are two important cases when this treatment cannot be used.

5.9.1 A conjugate problem

A conjugate problem is one in which two or more regions in the domain have distinct values for the diffusivity and there is a "jump" in the diffusivity across the interface between the regions, e.g., heat conduction in a liquid metal contained in a mold. If the location of the sub-regions are fixed and known, the best way to deal with this problem is to construct the mesh so that any given element falls in one and only one region, see Fig (5.6). In this way, since the CVFEM is constructed from the individual triangular elements there is no ambiguity in calculating the face diffusivities in a given element.

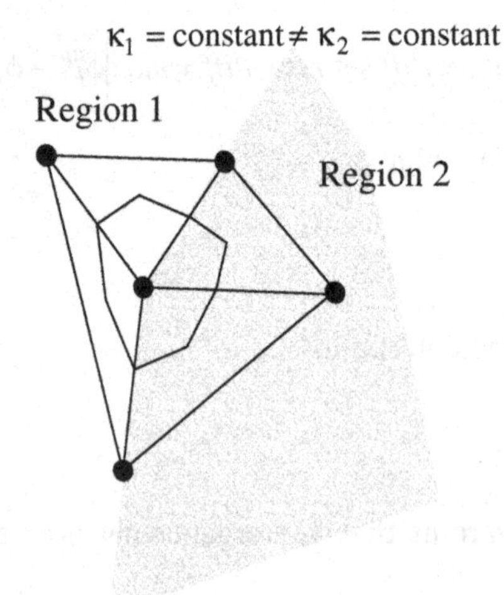

Fig. 5.6 Suggested mesh arrangement for a conjugate diffusion problem

5.9.2 Diffusivity a function of field variable

In many models and problems the diffusivity is a function of the unknown field, i.e., $\kappa = f(\phi)$. While the basic treatment outlined in §5.3 will work for this case, the resulting system of equations (5.55) will be non-linear. This will require an iterative solution with the full recalculation of the CVFEM diffusive coefficients at each step. An alternative, more efficient approach for this case is a follows. If the diffusive contributions to the coefficients are separated and stored separately, the discrete equation (5.55) can be rewritten as

$$[\, a_i^u + Q_{C_i} + B_{C_i} \,]\phi_i$$

$$= \sum_{j=1}^{n_i} a_{i,j}^k \phi_{S_{i,j}} - a_i^k \phi_i + \sum_{j=1}^{n_i} a_{i,j}^u \phi_{S_{i,j}} + Q_{B_i} + B_{B_i} \quad (5.61)$$

For a given element, on noting the properties of the shape function and its derivatives,

$1 - N_1 = N_2 + N_3 \Rightarrow -N_{1x} = N_{2x} + N_{3x}$ and $-N_{1y} = N_{2y} + N_{3y}$,

it follows from (5.27) that in an element, $a_1^k = a_2^k + a_3^k$ and that in a region of support,

$$a_i^k = \sum_{j=1}^{n_i} a_{i,j}^k$$

In this way, on first calculating the diffusion coefficients assuming $\kappa = 1$, a version of (5.61), accounting for the variable diffusivity, is given by

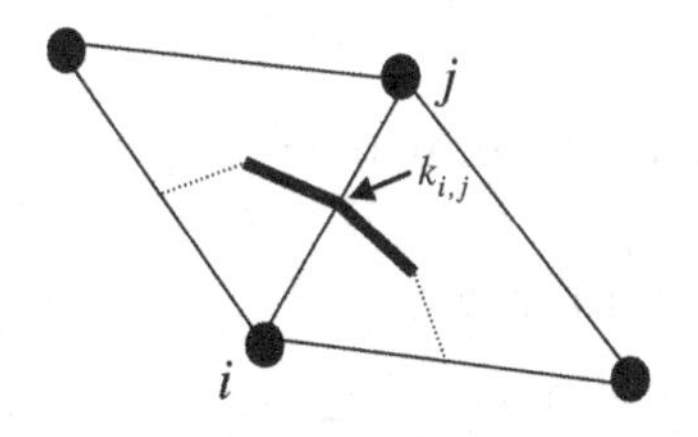

Fig. 5.7 Diffusivity at faces meeting on side *i-j*

$$[\, a_i^u + Q_{C_i} + B_{C_i} \,]\phi_i = \sum_{j=1}^{n_i} \kappa_{i,j} a_{i,j}^k (\phi_{S_{i,j}} - \phi_i) + \sum_{j=1}^{n_i} a_{i,j}^u \phi_{S_{i,j}} + Q_{B_i} + B_{B_i} \tag{5.62}$$

where the diffusivity acting on the two faces joining on the element side connecting node i and j, see Fig 5.7, is

$$\kappa_{i,j} = \frac{\kappa(\phi_i) + \kappa(\phi_{S_{i,j}})}{2} \tag{5.63}$$

Equation (5.62) can be readily manipulated into the iterative form

$$\phi_i^r = \phi_i^{r-1} + \omega \left(\frac{\left[\sum_{j=1}^{n_i} (a_{i,j}^u + \kappa_{i,j} a_{i,j}^k) \phi_{S_{i,j}} + Q_{B_i} + B_{B_i} \right]^{r,r-1}}{[\, a_i^u + \sum_{j=1}^{n_i} \kappa_{i,j} a_i^k + Q_{C_i} + B_{C_i} \,]} - \frac{[\, a_i^u + k_{i,j} a_{i,j}^k + Q_{C_i} + B_{C_i} \,]\phi_i^{r-1}}{[\, a_i^u + \sum_{j=1}^{n_i} \kappa_{i,j} a_i^k + Q_{C_i} + B_{C_i} \,]} \right) \tag{5.64}$$

Although this is more involved than the basic iterative scheme in (5.60) the additional computations in forming this equation during the iteration are minimal. At node i during the rth iteration, perform the following steps

----Loop over the nodes, $j = 1..n_i$, in the ith region of support

------**Calculate $\kappa_{i,j}$ from (5.63)

------Calculate and store the value $\sum_{j=1}^{n_i} (a_{i,j}^u + \kappa_{i,j} a_{i,j}^k) \phi_{S_{i,j}}$

------**Calculate the sum $\sum_{j=1}^{n_i} \kappa_{i,j}$

----Close Loop on j

The only additional calculations from the basic iteration are marked with (**). This additional calculation is much less intensive that the alternative of reforming the diffusive coefficients on each iterative step.

5.10 Transients

From the above the net flow rate of quantity *into* the control volume around node i can be approximated as

$$Net_i = \int_{V_i} Q\, dV + \sum_{j=1}^{n_i} \int_{A^j} \kappa \nabla\phi \cdot \boldsymbol{n}\, dA - \int_{A^j} (\boldsymbol{v} \cdot \boldsymbol{n})\phi\, dA$$

$$\approx -[\, a_i + Q_{C_i} + B_{C_i}\,]\phi_i + \sum_{j=1}^{n_i} a_{i,j}\phi_{S_{i,j}} + Q_{B_i} + B_{B_i} \qquad (5.65)$$

In a steady state problem this flow rate will be identically zero. In a transient problem, however, it will result in a change in storage of the quantity in the control volume, i.e.,

$$\frac{d}{dt} \int_{V_i} \phi dV = Net_i \qquad (5.66)$$

Using nodal lumping for the volume integration and finite difference in time this equation can be used to evaluate the nodal field values at time $t + \Delta t$ in terms of the nodal field values at time t,

$$V_i \phi_i^{new} = V_i \phi_i + \Delta t [\, (1-\theta) Net_i + \theta Net_i^{new}\,] \qquad (5.67)$$

where Δt is a time step and the superscript $()^{new}$ indicates evaluation at time $t + \Delta t$. The parameter $0 \le \theta \le 1$ is a user-defined weighting factor, used to approximate the net flow into control volume i during the time interval $[t, t+\Delta t]$, in terms of the net flows at the beginning and end of the time step. Neglecting, for now, contributions from the boundaries and sources, the resulting discrete equations for three choices of θ are:

Fully Implicit $\theta = 1$:

$$V_i \phi_i^{new} = V_i \phi_i + \Delta t \left(\sum_{j=1}^{n_i} a_{i,j} \phi_{S_{i,j}}^{new} - a_i \phi_i^{new} \right) \tag{5.68}$$

The advantage of this choice is that it is unconditionally stable—i.e., for any choice of time step errors (induced or inherent) will not grow. The downside is that a system of equations needs to be solved to obtain the field values at the new time step $\phi_i^{new}\,(i = 1..n)$.

Crank-Nicolson $\theta = .5$:

$$V_i \phi_i^{new} = V_i \phi_i + \frac{\Delta t}{2} \left(\sum_{j=1}^{n_i} a_{i,j} \phi_{S_{i,j}}^{new} - a_i \phi_i^{new} \right) + \frac{\Delta t}{2} \left(\sum_{j=1}^{n_i} a_{i,j} \phi_{S_{i,j}} - a_i \phi_i \right) \tag{5.69}$$

This un-conditionally stable scheme also requires a system solution. Unlike the 1st order in time, fully implicit method, however, the Crank-Nicolson is 2nd order in time (i.e., time errors scale with Δt^2. By the way there is no h in Nicolson, see Crank and Nicolson (1947).

Fully Explicit $\theta = 0$:

$$V_i \phi_i^{new} = V_i \phi_i + \Delta t \left(\sum_{j=1}^{n_i} a_{i,j} \phi_{S_{i,j}} - a_i \phi_i \right) \tag{5.70}$$

This choice can be used to directly update the new time level values from the current time values *without* solution of a system of equations. This, however, comes at the price of a restriction on the time step size to ensure stability. The solution of (5.70) is likely to become unstable (error grow as the solution advances) if the net coefficient for ϕ_i becomes negative. This requires that the time step is chosen such that

$$\Delta t < min\left(\frac{V_i}{a_i}\right), \quad (i = ..n) \tag{5.71}$$

If a fine grid is used this value could be prohibitively small, i.e., the advance of time could be too slow to reach a practical value within reasonable computational resources. In many cases, however, the drawback for using a small time step is offset by the ability to solve (5.70) without iteration. Driven by its simplicity and flexibility of modeling complex non-linear terms the explicit time integration approach is the preferred choice in this work.

When sources and boundary condition treatments are added the explicit scheme (5.70) can be written as

$$(V_i + B_{C_i})\phi_i^{new} = V_i\phi_i + \Delta t\left(\sum_{j=1}^{n_i} a_{i,j}\phi_{S_{i,j}} - a_i\phi_i + Q_{B_i} - Q_{C_i}\phi_i\right) + B_{B_i} \tag{5.72}$$

The addition of the source term in (5.72) could require a further reduction in the time step to retain positive coefficients and stability, i.e,.

$$\Delta t < min\left(\frac{V_i}{a_i + Q_{C_i}}\right), \quad (i = ..n)$$

Further, the boundary treatment in (5.72) allows for an insulated boundary on setting $B_{C_i} = 0$ and $B_{B_i} = 0$ and a fixed value boundary on setting $B_{C_i} = 10^{16}$ and $B_{B_i} = B_{C_i}\phi_{value}$.

The coding for one time step of the explicit scheme is very straightforward.

The coefficients, source terms, and boundary treatments (fixed or insulated) are all calculated by the steps outlined previously.
An appropriate time step is chosen.

++A loop is made over the domain nodes $i = 1..n$

-- Looping over the nodes, $j = 1..n_i$, in the ith region of support

---- The value $\sum_{j=1}^{n_i} a_{i,j}\phi_{S_{i,j}}$ of is calculated and stored.

-- Close loop on *j*

-- The update new time value at node *i* is found directly from (5.70)

++Close loop on *i*

[**note** In solving over multiple time steps up to the simulation end time, *maxtime*, the loop above, denoted by ++, is inserted in an outer time loop]

$itim = 1..maxtim$

++

++

Close loop on *itim*

5.11 Summary

At this point the key steps in a CVFEM discretization and solution of the advection-diffusion equation (ADE) have been provided. The differences between the CVFEM treatment of the ADE and treatments for solid and fluid mechanics problems are essentially in the details. Hence, the information provided in this chapter also enables the CVFEM treatments of solid and fluids problems.

Chapter 6

The Control Volume Finite Difference Method

To provide a contrast with the CVFEM a brief overview of control volume finite difference methods (CVFDM) is provided.

6.1 The Task

At this point it is very worthwhile to provide a brief overview of control volume finite difference methods (CVFDM). This serves two purposes:

1. It reinforces some of the key concepts of CVFEM

2. It provides an alternative to CVFEM; an alternative that in some cases may prove to be an easier and more convenient solution approach for the problem at hand

In the presentation of the CVFDM we will focus on the two-dimensional steady state advection diffusion equation, in integral form

$$\int_V Q \; dV + \int_A \kappa \nabla \phi \cdot \boldsymbol{n} \, dA - \int_A (\boldsymbol{v} \cdot \boldsymbol{n}) \phi \, dA = 0 \qquad (6.1)$$

The central task is to write this equation in a discrete form—similar in nature to the CVFEM discrete equation (6.55).

6.2 CVFDM Data Structure

The key components of the data structure for a CVFDM discretization are shown in Fig 6.1. Note, *for the case shown*

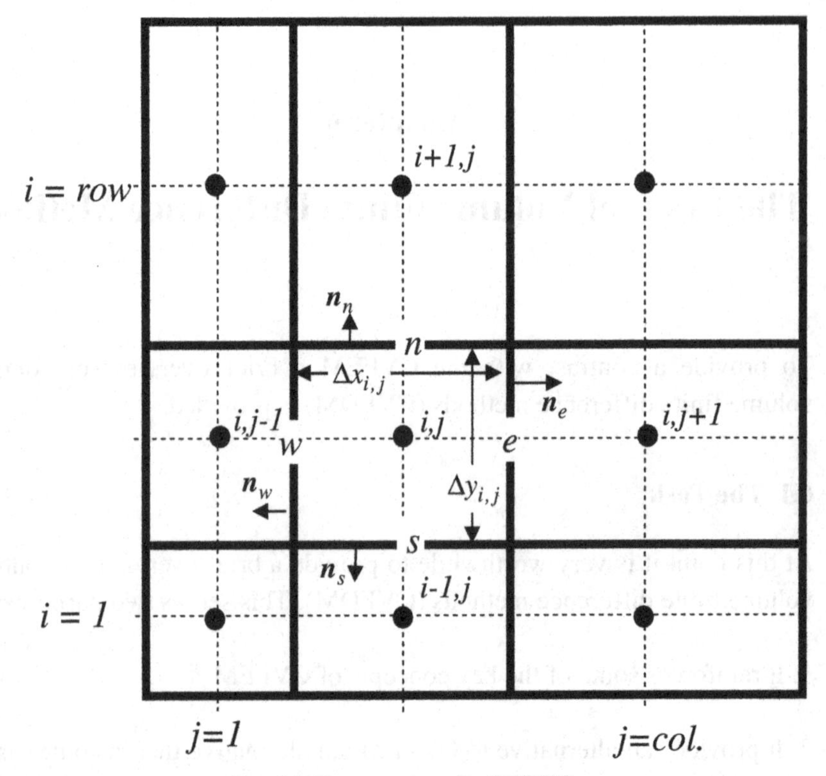

Fig. 6.1 Data structure for CVFDM

1. the rectangular control volumes (bold lines) fit the solution domain,

2. the nodes, located at the control volume centers, are on a structured mesh (dashed lines), i.e., a node location can be located with a row (*i*) column (*j*) index $()_{i,j}$,

3. the faces of the control volume around node $()_{i,j}$ are labeled *n*(orth), *w*(est), *s*(outh), and *e*(ast),

4. the unit normal on the control volume faces are

$$\boldsymbol{n}_n = (0,1);\ \boldsymbol{n}_w = (-1,0);\ \boldsymbol{n}_s = (-1,0);\ \boldsymbol{n}_e = (1,0)$$

5. the areas of these faces (assuming unit depth) are

$$A_n = A_s = \Delta y_{i,j},\quad A_e = A_w = \Delta x_{i,j}$$

6. the volume of the control volume around node $()_{i,j}$ is

$$V_{i,j} = \Delta x_{i,j} \times \Delta y_{i,j}$$

7. the distance between node $()_{i,j}$ and its neighbors to the to the north, west, south and east are, respectively

$$\delta_n = \frac{\Delta y_{i,j} + \Delta y_{i+1,j}}{2}, \quad \delta_w = \frac{\Delta x_{i,j} + \Delta x_{i,j-1}}{2}$$

$$\delta_s = \frac{\Delta y_{i,j} + \Delta y_{i-1,j}}{2}, \quad \delta_e = \frac{\Delta x_{i,j} + \Delta x_{i,j+1}}{2}$$

In nodes adjacent to domain boundaries these values measure the distance from the node to the respective boundary.

With the data structure of Fig 6.1 the discrete equation for (6.1) will take the form

$$[\, a_{P_{ij}} + Q_{C_{ij}} \,]\phi_{i,j} = a_{n_{ij}}\phi_{i+1,j} + a_{w_{ij}}\phi_{i,j-1} + a_{s_{ij}}\phi_{i-1,j} + a_{e_{ij}}\phi_{i,j+1} + Q_{B_{ij}} + B_{ij} \tag{6.2}$$

i.e., for each node $()_{i,j}$ in the domain, 5 coefficients $(\, a_p, a_n, a_w, a_s, a_e \,)$, associated with the node and its neighbors, linearized source terms $(\, Q_{C_{ij}}$ and $Q_{B_{ij}} \,)$, and a single boundary term B_{ij} will be identified.

6.3 Coefficients and Sources

When the domains of integration in (6.1) are associated with the node $()_{i,j}$ the governing equation can be written as

$$\begin{aligned} &\int_V Q \; dV + \int_A \kappa\nabla\phi\cdot\boldsymbol{n}\,dA - \int_A (\boldsymbol{v}\cdot\boldsymbol{n})\phi\,dA = \\ &\qquad \int_{V_{i,j}} Q \; dV + \sum_{k=n,w,s,e} \int_{A_k} \kappa\nabla\phi\cdot\boldsymbol{n}\,dA - \int_{A_k} (\boldsymbol{v}\cdot\boldsymbol{n})\phi\,dA \end{aligned} \tag{6.3}$$

Employing midpoint integration rules the right hand side of (6.3) can be approximated by

$$\int_{V_{i,j}} Q\,dV + \sum_{k=n,w,s,e} \int_{A_k} \kappa\nabla\phi\cdot\boldsymbol{n}\,dA - \int_{A_k}(\boldsymbol{v}\cdot\boldsymbol{n})\phi\,dA \approx$$
$$Q_{i,j}V_{i,j} + \sum_{k=n,w,s,e} A_k[\kappa\nabla\phi\cdot\boldsymbol{n}]_k - \sum_{k=n,w,s,e} A_k[(\boldsymbol{v}\cdot\boldsymbol{n})\phi]_k \tag{6.4}$$

where $[]_k$ indicates evaluation at the midpoint of the face $k = n,w,s$, or e. On using source linearization (assuming a volume source)

$$Q_{i,j}V_{i,j} = -Q_{C_{ij}}\phi_{i,j} + Q_{B_{ij}}, \qquad Q_{C_{ij}}, Q_{B_{ij}} \geq 0 \tag{6.5}$$

and identifying

$$q_k = A_k[(\boldsymbol{v}\cdot\boldsymbol{n})]_k \tag{6.6}$$

as the volume flow out across the kth face we arrive at the following discrete form of (6.1)

$$-Q_{C_{i,j}}\phi_{i,j} + Q_{B_{i,j}} + \sum_{k=n,w,s,e} A_k[\kappa\nabla\phi\cdot\boldsymbol{n}]_k - \sum_{k=n,w,s,e} q_k\phi_k = 0 \tag{6.7}$$

Further refinement can be made by considering the diffusive and advective fluxes across the control volume faces. On noting that the dot product

$$[\nabla\phi]\cdot\boldsymbol{n} = \frac{\partial\phi}{\partial x}n_x + \frac{\partial\phi}{\partial y}n_y \tag{6.8}$$

approximating face diffusivities as distance-weighed averages of neighboring nodal values,

$$\kappa_n = \frac{\Delta y_{i+1,j}\kappa_{i,j} + \Delta y_{i,j}\kappa_{i+1,j}}{\Delta y_{i,j} + \Delta y_{i+1,j}}, \quad \kappa_w = \frac{\Delta x_{i,j-1}\kappa_{i,j} + \Delta x_{i,j}\kappa_{i,j-1}}{\Delta x_{i,j} + \Delta x_{i,j-1}},$$
$$\kappa_s = \frac{\Delta y_{i-1,j}\kappa_{i,j} + \Delta y_{i,j}\kappa_{i-1,j}}{\Delta y_{i,j} + \Delta y_{i-1,j}}, \quad \kappa_e = \frac{\Delta x_{i,j+1}\kappa_{i,j} + \Delta x_{i,j}\kappa_{i,j+1}}{\Delta x_{i,j} + \Delta x_{i,j+1}} \tag{6.9}$$

and using a central finite differences to approximate face derivatives the diffusive fluxes in (6.7) can be written as

$$
\begin{aligned}
A_n[\kappa\nabla\phi]_n \cdot \boldsymbol{n}_n &= \kappa_n A_n \left.\frac{\partial\phi}{\partial y}\right|_n = \kappa_n \Delta x_{i,j} \frac{[\phi_{i+1,j} - \phi_{i,j}]}{\delta_n} \\
A_w[\kappa\nabla\phi]_w \cdot \boldsymbol{n}_w &= -\kappa_w A_w \left.\frac{\partial\phi}{\partial x}\right|_w = \kappa_w \Delta y_{i,j} \frac{[\phi_{i,j-1} - \phi_{i,j}]}{\delta_w} \\
A_s[\kappa\nabla\phi]_s \cdot \boldsymbol{n}_s &= -\kappa_s A_s \left.\frac{\partial\phi}{\partial y}\right|_s = \kappa_s \Delta x_{i,j} \frac{[\phi_{i-1,j} - \phi_{i,j}]}{\delta_s} \\
A_e[\kappa\nabla\phi]_e \cdot \boldsymbol{n}_e &= \kappa_e A_e \left.\frac{\partial\phi}{\partial x}\right|_e = \kappa_e \Delta y_{i,j} \frac{[\phi_{i,j+1} - \phi_{i,j}]}{\delta_e}
\end{aligned}
\tag{6.10}
$$

Further, on approximating the velocity components, v_x and v_y, as distance-weighed averages of neighboring nodal values,

$$
\begin{aligned}
v_{y_n} &= \frac{\Delta y_{i+1,j} v_{y_{i,j}} + \Delta y_{i,j} v_{y_{i+1,j}}}{\Delta y_{i,j} + \Delta y_{i+1,j}}, \quad v_{x_w} = \frac{\Delta x_{i,j-1} v_{x_{i,j}} + \Delta x_{i,j} v_{x_{i,j-1}}}{\Delta x_{i,j} + \Delta x_{i,j-1}} \\
v_{y_s} &= \frac{\Delta y_{i-1,j} v_{y_{i,j}} + \Delta y_{i,j} v_{y_{i-1,j}}}{\Delta y_{i,j} + \Delta y_{i-1,j}}, \quad v_{x_e} = \frac{\Delta x_{i,j+1} v_{x_{i,j}} + \Delta x_{i,j} v_{x_{i,j+1}}}{\Delta x_{i,j} + \Delta x_{i,j+1}}
\end{aligned}
\tag{6.11}
$$

the volume flows across the control volume faces can be evaluated by

$$
q_n = \Delta x_{i,j} v_{y_n}, \quad q_w = -\Delta y_{i,j} v_{x_w}, \quad q_s = -\Delta x_{i,j} v_{y_s}, \quad q_e = \Delta y_{i,j} v_{x_e} \tag{6.12}
$$

In this way, on using upwinding to set the face value in the advective term according to the flow direction, the advective fluxes in (6.7) can be written as

$$
\begin{aligned}
q_n\phi_n &= max(0, q_n)\phi_{i,j} - max(0, -q_n)\phi_{i+1,j} \\
q_w\phi_w &= max(0, q_w)\phi_{i,j} - max(0, -q_w)\phi_{i,j-1} \\
q_s\phi_s &= max(0, q_s)\phi_{i,j} - max(0, -q_s)\phi_{i-1,j} \\
q_e\phi_e &= max(0, q_e)\phi_{i,j} - max(0, -q_e)\phi_{i,j+1}
\end{aligned}
\tag{6.13}
$$

Note that continuity requires

$$
\sum_{k=n,w,s,e} max(0, q_k) - max(0 - q_k) = 0 \tag{6.14}
$$

On substituting (6.13) and (6.10) in (6.4), gathering terms and using (6.14), the coefficients in (6.2) can be readily identified

$$a_{n_{ij}} = \kappa_n \frac{\Delta x_{i,j}}{\delta_n} + max(0 - q_n), \quad a_{w_{ij}} = \kappa_w \frac{\Delta y_{i,j}}{\delta_w} + max(0 - q_w)$$

$$a_{s_{ij}} = \kappa_n \frac{\Delta x_{i,j}}{\delta_s} + max(0 - q_s), \quad a_{e_{ij}} = \kappa_e \frac{\Delta y_{i,j}}{\delta_e} + max(0 - q_e) \qquad (6.15)$$

$$a_{p_{ij}} = \sum_{k=n,w,s,e} a_{k_{ij}}$$

6.4 Boundary Conditions

Equation (6.7) and (6.15) respectively define the source terms and coefficients in the CVFDM discrete form, (6.7), of the steady state advection diffusion equation (6.1). To fully define the equation, however, specification of the source term B_{ij} needs to be made. The default setting for this term is $B_{ij} = 0$. A finite value is only required at node points located in control volumes that share at least one face with a domain boundary.

6.4.1 Insulated (no-flow) boundary

If the domain boundary is an insulated (no-flow) boundary the default value $B_{ij} = 0$ is retained. In addition the control volume face coefficient associated with the domain boundary is also set to zero. For example, if the NORTH domain boundary is an insulated boundary the coefficient setting $a_{n_{ij}} = 0$ is made; note this setting *must* be reflected in the calculation of the node point coefficient

$$a_{p_{ij}} = \sum_{k=w,s,e} a_{k_{ij}} \qquad (6.16)$$

6.4.2 Fixed value boundary

If the domain boundary (NORTH say) is a fixed value boundary

$\phi = \phi_{value}$ the following coding can be used to ensure a correct setting in (6.2):

1. Calculate the north coefficient from (6.15), i.e.,

$$a_{n_{ij}} = \kappa_n \frac{\Delta x_{i,j}}{\delta_n} + max(0 - q_n)$$

Take care to calculate δ_n as the distance from the node to the boundary.

2. Set $B_{ij} = a_{n_{ij}} \phi_{value}$.
3. Reset $a_{n_{ij}} = 0$ and update $a_{p_{ij}}$ through (6.16).

6.5 Summary

The object here has been to provide sufficient information to allow readers the opportunity to use the CVFDM in place of the CVFEM. Even though the CVFDM has only been developed for the case of steady state advection-diffusion, it is hoped that the reader is able to recognize the extensive similarities between the CVFDM and CVFEM and thereby adapt the more general treatment of CVFEM presented in Chapter 5 to CVFDM.

Chapter 7

Analytical and CVFEM Solutions of Advection-Diffusion Equations

The control volume finite element method (CVFEM) is applied to a wide range of advection-diffusion equations (ADE's). In every case there is an analytical solution available such that a rigorous test of the performance of the CVFEM can be made.

7.1 The Task

Up to this point the essential ingredients for the CVFEM solution of the transient, two-dimensional advection-diffusion equation (ADE) have been detailed. The key result has been to show how the ADE in integral form

$$\frac{d}{dt}\int_V \phi dV - \int_V Q \; dV - \int_A \kappa\nabla\phi\cdot\boldsymbol{n}\, dA + \int_A (\boldsymbol{v}\cdot\boldsymbol{n})\phi\, dA = 0 \tag{7.1}$$

or point form

$$\frac{\partial\phi}{\partial t} + \nabla\cdot(\boldsymbol{v}\phi) - \nabla\cdot(\kappa\nabla\phi) - Q = 0 \tag{7.2}$$

can be represented by the system of CVFEM discrete equations (5.55)

$$[\, a_i + Q_{C_i} + B_{C_i} \,]\phi_i = \sum_{j=1}^{n_i} a_{i,j}\phi_{S_{i,j}} + Q_{B_i} + B_{B_i} \tag{7.3}$$

in steady state problems, or (5.72)

$$(V_i + B_{C_i})\phi_i^{new} = V_i\phi_i + \Delta t\left(\sum_{j=1}^{n_i} a_{i,j}\phi_{S_{i,j}} - a_i\phi_i + Q_{B_i} - Q_{C_i}\phi_i\right) + B_{B_i} \tag{7.4}$$

in transient problems; where the nodal values ϕ_i and $\phi_{S_{i,j}}$ are located at the vertices of an unstructured mesh of linear triangular finite elements. The task at hand is to test and validate solutions of ADE's based on these discretizations. Access to test advection diffusion problems that admit analytical solutions in is invaluable in fulfilling this task. Such problems not only provide a measure of the accuracy of the solution scheme developed but also provide a "safety net" that prevents conceptual and coding errors.

7.2 Choice of Test Problems

In Cartesian coordinate systems, while there are a variety of solutions available for the one-dimensional version of (7.2)

$$\frac{\partial \phi}{\partial t} + \frac{\partial}{\partial x}(v_x \phi) - \frac{\partial}{\partial x}\left(\kappa \frac{\partial \phi}{\partial x}\right) - Q = 0, \tag{7.5}$$

there are no two-dimensional Cartesian solutions that include all of the terms in (7.2). One-dimensional solutions are a useful start but are limited in their ability to fully test CVFEM solutions of (7.2). It is possible, however, to find a full range of solutions for the axisymmetric form of (7.2)

$$\frac{\partial \phi}{\partial t} + \frac{1}{r}\frac{\partial}{\partial r}(r v_r \phi) = \frac{1}{r}\frac{\partial}{\partial r}\left(r\kappa \frac{\partial \phi}{\partial r}\right) + Q \tag{7.6}$$

where $v_r \sim \frac{1}{r}$ is now the radial velocity. At first site, since problems governed by (7.6) are essentially one-dimensional, these solutions seem to offer no more advantage than the Cartesian one-dimensional solutions of (7.2). Many of the one-dimensional axisymmetric problems that admit analytical solutions, however, can be cast as two-dimensional Cartesian problems. In the Cartesian framework these two-dimensional problems provide a full test for the multi-dimensional capabilities of the CVFEM, and allow for exact performance measures on comparing the Cartesian numerical prediction with the axisymmetric analytical values. In this chapter the analytical solution of problems described by the advection-diffusion equation of the forms (7.5) and (7.6) will be presented and then extensively compared with CVFEM solutions operating in two-dimensional Cartesian space. Basic geometries for these problems are shown in Fig. 7.1 (one-dimensional domains and two-dimensional equivalents) and Fig. 7.2 (polar coordinates and Cartesian equivalents).

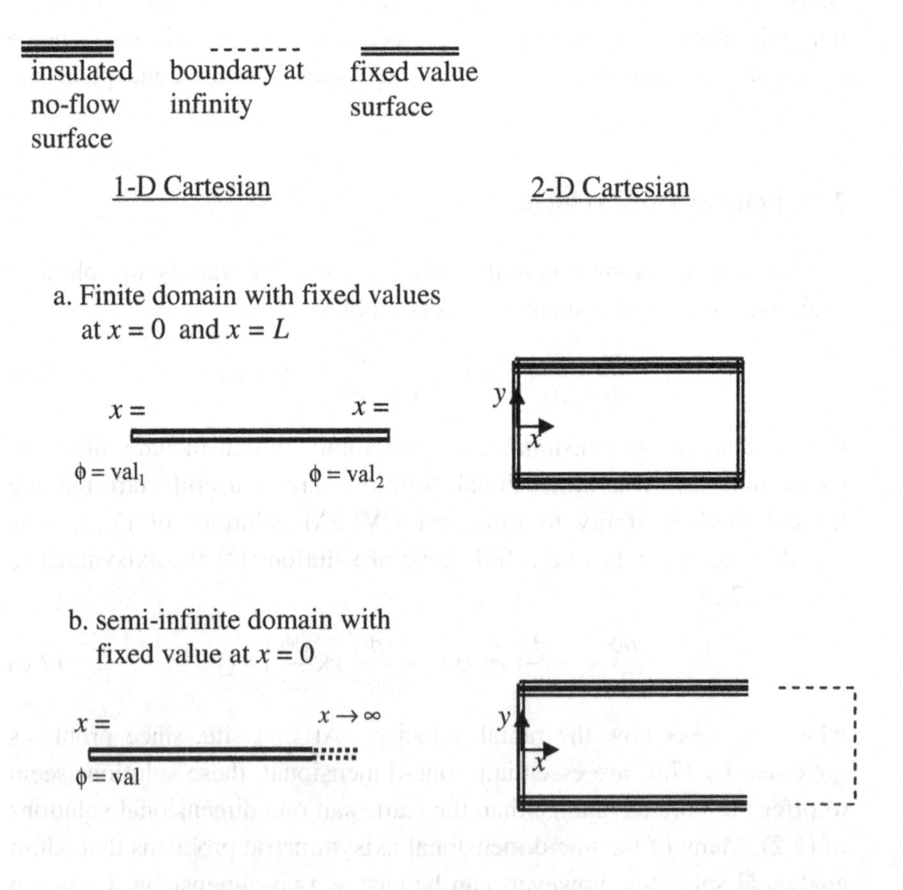

Fig. 7.1 One-dimensional domains and two-dimensional equivalents for one-dimensional advection-diffusion problems

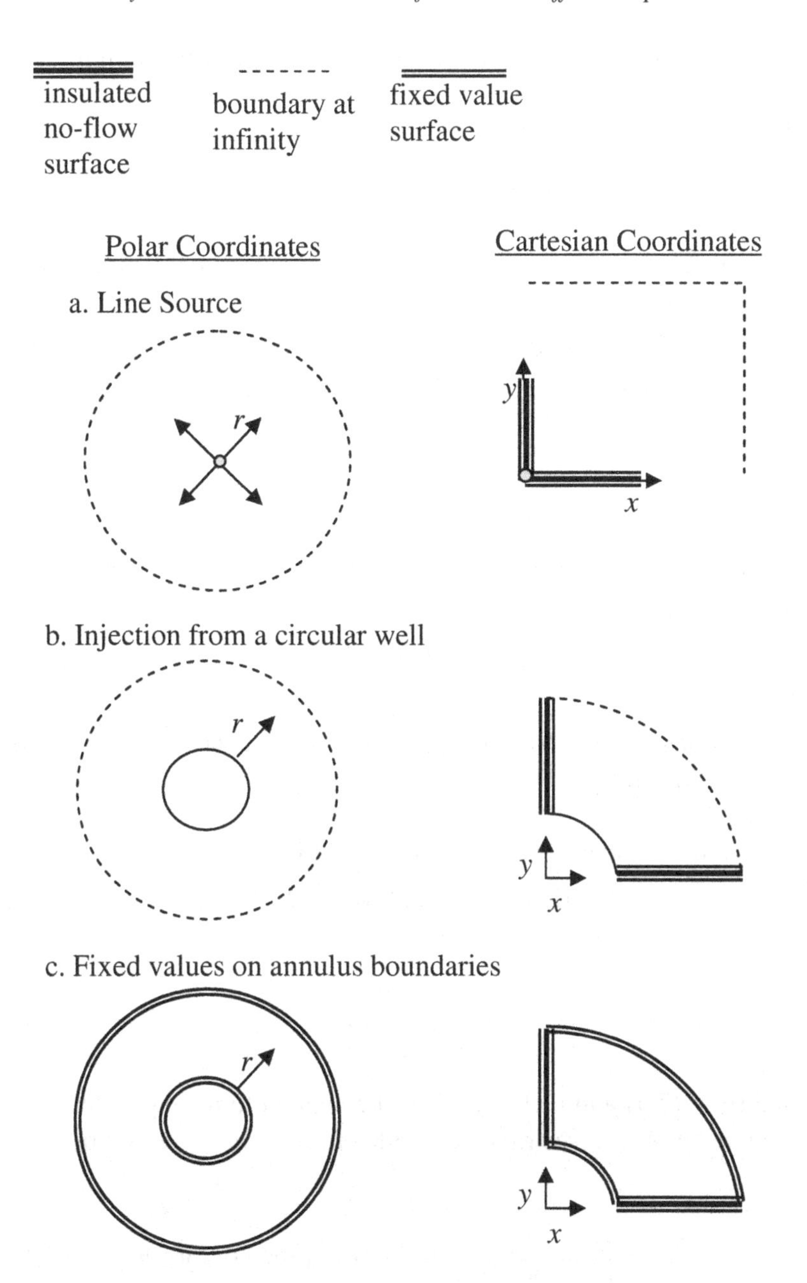

Fig. 7.2 One-dimensional domains and two-dimensional equivalents for axi-symmetric advection-diffusion problems

7.3 One-Dimensional Steady State Diffusion in a Finite Domain

For this problem, (7.5) has the form

$$\frac{d}{dx}\left(\kappa\frac{d\phi}{dx}\right)=0,\quad 0\leq x\leq L \tag{7.7}$$

in the fixed domain shown in Fig. 7.1a. If Dirichlet (fixed-value) boundary conditions are imposed

$$\phi(0)=\phi_1 \quad \text{and} \quad \phi(L)=\phi_2 \tag{7.8}$$

and a constant diffusivity κ is assumed, the solution of (7.7) satisfying (7.8) is

$$\phi=(\phi_2-\phi_1)\frac{x}{L}+\phi_1 \tag{7.9}$$

If the diffusivity is given by $\kappa=\alpha(x+1)$ where α is a constant the solution of (7.7) satisfying (7.8) is

$$\phi=(\phi_2-\phi_1)\frac{ln(x+1)}{ln(L+1)}+\phi_1 \tag{7.10}$$

The solutions of (7.9) and (7.10) are plotted for the case $\alpha=1, L=1,\ \phi_1=1, \phi_2=0$ in Fig. 7.3.

When the boundary conditions are given by

$$\phi(x=0,y)=\phi_1 \quad \text{and} \quad \phi(x=L,y)=\phi_2 \tag{7.11a}$$

and

$$\left.\frac{\partial\phi}{\partial y}\right|_{y=\pm b}=0 \tag{7.11b}$$

the solutions (7.9) and (7.10) can also be applied to the equivalent two-dimensional problem (see right hand side of Fig. 7.1a) governed by

$$\frac{\partial}{\partial x}\left(\kappa\frac{\partial\phi}{\partial x}\right)+\frac{\partial}{\partial y}\left(\kappa\frac{\partial\phi}{\partial y}\right)=0,\quad 0\leq x\leq L,\ 0\leq y\leq b \tag{7.12}$$

where b is an arbitrary constant.

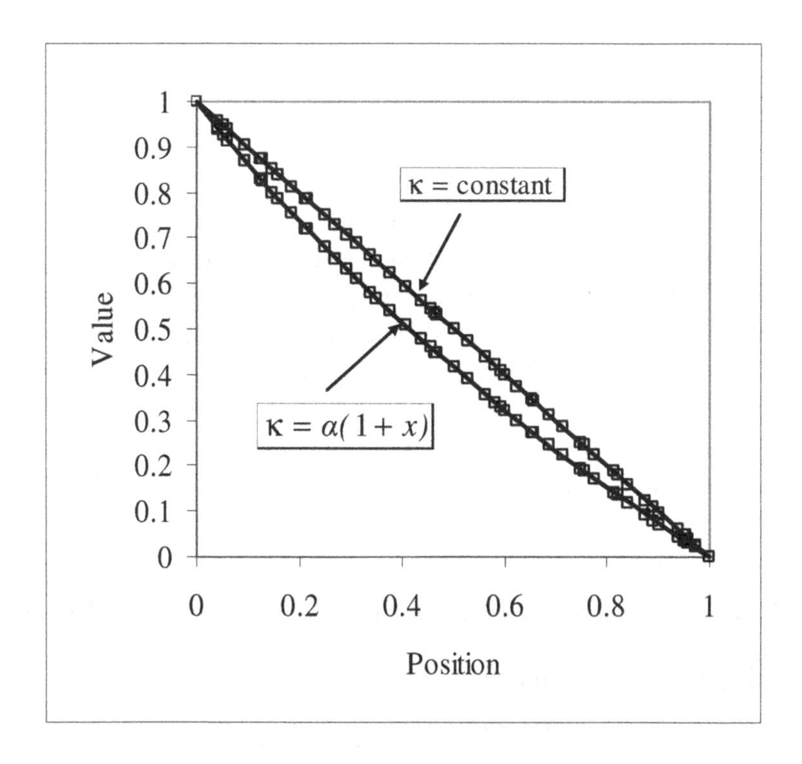

Fig. 7.3 Analytical (solid lines) and CVFEM (open squares) solutions of (7.12)

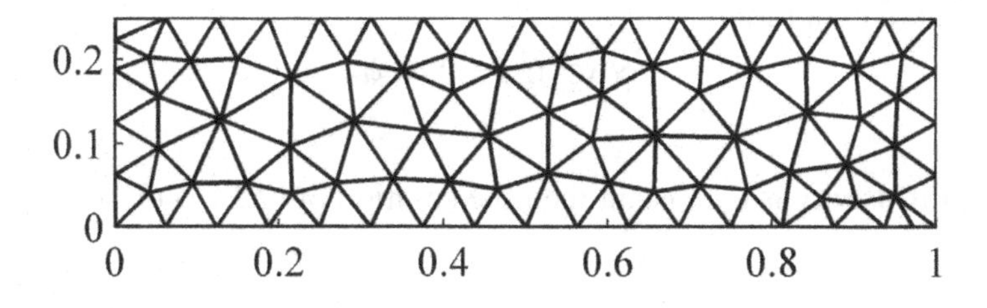

Fig. 7.4 Mesh for CVFEM solution of (7.12)

The CVFEM discretization for this problem, implemented on the mesh shown in Fig. 7.4, is

$$[\,a_i + B_{C_i}\,]\phi_i = \sum_{j=1}^{n_i} a_{i,j}\phi_{S_{i,j}} + B_{B_i} \tag{7.13}$$

The coefficients are generated from (5.27) and the boundary terms are set such that (i) the upper (y = b= 0.25) and lower (y = 0) are insulated ($B_C = 0, B_B = 0$), (ii) the left hand boundary (x = 0) is the fixed value 1 ($B_C = 10^{16}, B_B = 10^{16}$), and (iii) the right hand boundary (x = 1) is the fixed value 0 ($B_C = 10^{16}, B_B = 0$).The CVFEM solutions, obtained from a point wise iteration of (7.13) are shown as the open squares superposed on the analytical lines in Fig. 7.3. Although somewhat trivial the agreement between the CVFEM and the analytical solution is excellent. Note that this problem is a reasonable first test for the CVFEM, since the grid in Fig. 7.4 is most defiantly not one-dimensional.

7.4 One-Dimensional Transient Diffusion in a Semi-Infinite Domain

For this problem (7.5) has the form

$$\frac{\partial\phi}{\partial t} - \frac{\partial}{\partial x}\left(\kappa\frac{\partial\phi}{\partial x}\right) = 0, \quad 0 \le x \tag{7.14}$$

in the semi-infinite domain of Fig. 7.1b. Suitable boundary conditions are

$$\phi(\,x = 0, t > 0\,) = \phi_1 \text{ and } \lim_{x\to\infty} \phi = \phi_2 \tag{7.15}$$

with initial condition

$$\phi(\,x > 0, t = 0\,) = \phi_2 \tag{7.16}$$

If the diffusivity is a constant then the solution of (7.14) - (7.16) is given by

$$\phi = (\,\phi_2 - \phi_1\,)\,erf\left(\frac{x}{2\sqrt{\kappa t}}\right) + \phi_1 \tag{7.17}$$

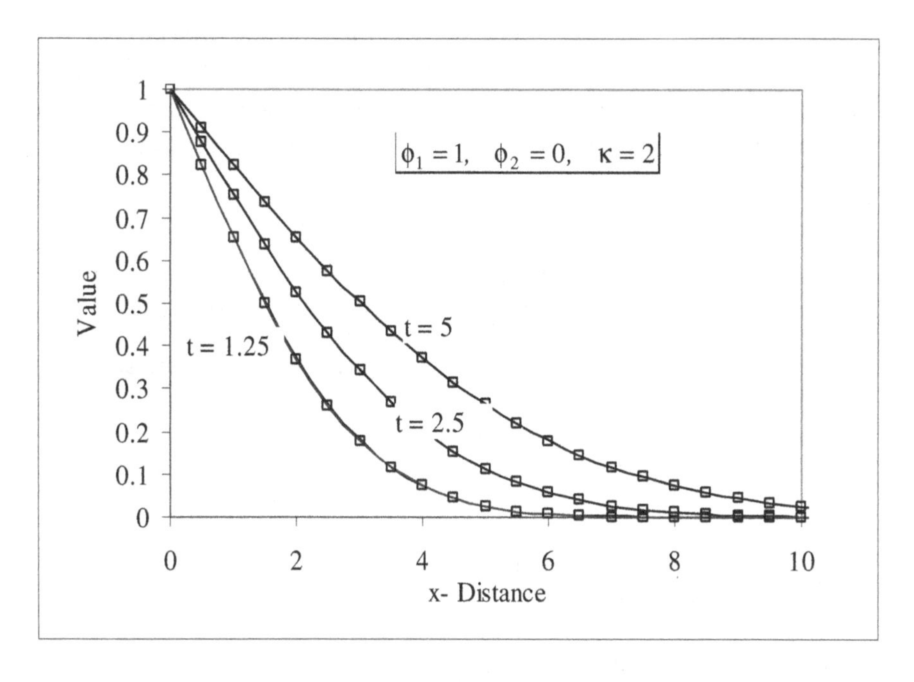

Fig. 7.5 Analytical (solid lines) and CVFEM (open squares) solutions of (7.22)

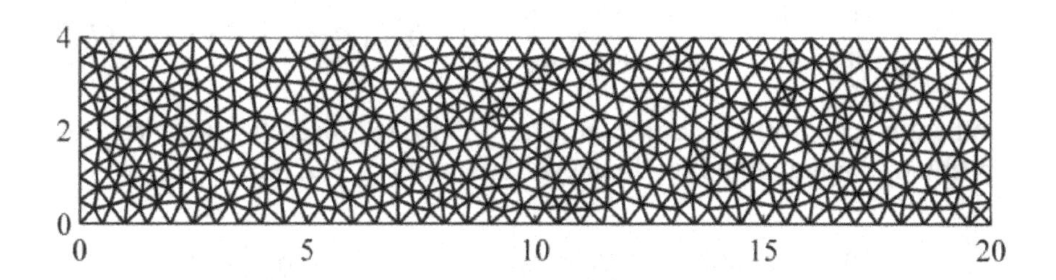

Fig. 7.6 Mesh for CVFEM solution of (7.22)

where

$$erf(z) = \frac{2}{\sqrt{\pi}} \int_0^{\alpha} e^{-\alpha^2} d\alpha \tag{7.18}$$

is the error function (page 297 in Abramowitz and Stegun (1970)). Calculation of the error function is built into many available commercial computational packages. If a calculator is not available, however, it can be estimated, to within an absolute error of 5×10^{-4}, from the rational approximation

$$erf(x)=1-\frac{1}{[1+a_1x+a_2x^2+a_3x^3+a_4x^4]^4} \quad (7.19)$$

$$a_1=.278393,\ a_2=.230389, a_3=.000972, a_4=.078108$$

Example plots of the solution in (7.17), at three time steps, are shown as solid lines in Fig. 7.5.

On imposing the boundary conditions

$$\phi(x=0,-b\le y\le b,t>0)=\phi_1 \text{ and } \lim_{x\to\infty}\phi=\phi_2 \quad (7.20a)$$

$$\left.\frac{\partial\phi}{\partial y}\right|_{y=\pm b,\,0\le x<\infty}=0 \quad (7.20b)$$

and the initial condition

$$\phi(x>0,0\le y\le b,t=0)=\phi_2 \quad (7.21)$$

The solution (7.17) is also applicable to the two-dimensional problem

$$\frac{\partial\phi}{\partial t}-\frac{\partial}{\partial x}\left(\kappa\frac{\partial\phi}{\partial x}\right)-\frac{\partial}{\partial y}\left(\kappa\frac{\partial\phi}{\partial y}\right)=0,\quad 0\le x\le\infty,\quad 0\le y\le b \quad (7.22)$$

Where b is an arbitrary constant .The CVFEM discretization for this problem, implemented on the mesh shown in Fig 7.6, is

$$(V_i+B_{C_i})\phi_i^{new}=V_i\phi_i+\Delta t\left(\sum_{j=1}^{n_i}a_{i,j}\phi_{S_{i,j}}-a_i\phi_i\right)+B_{B_i} \quad (7.23)$$

In this mesh the long length (L = 20), chosen to mimic the far field $x\to\infty$ condition. The coefficients in (7.23) are generated from (5.27) and the boundary terms are set so that (i) the upper (y = b= 4), lower (y = 0) and right (far field) boundaries are insulated ($B_C=0, B_B=0$) and (ii) the left hand boundary (x = 0) is the fixed value $1(B_C=10^{16}, B_B=10^{16})$. The CVFEM solutions, obtained from the explicit time marching solution of (7.23) (profile along y = 0) with time step $\Delta t = 7.8125\times10^{-3}$ are shown as the open squares superposed on the analytical lines in Fig. 7.5. This solution is an excellent first test of the performance of control volume methods on transient problems.

7.5 One-dimensional Transient Advection-Diffusion in a Semi-Infinite Domain

Here all terms but the source term appears in the governing equation

$$\frac{\partial\phi}{\partial t}+\frac{\partial}{\partial x}(v_x\phi)-\frac{\partial}{\partial x}\left(\kappa\frac{\partial\phi}{\partial x}\right)=0,\quad 0\le x \tag{7.24}$$

Since the domain is one-dimensional (see Fig. 7.1b) the velocity v is a constant scalar. If the diffusivity is also a constant (7.14) can be written in the dimensionless form

$$\frac{\partial\phi}{\partial t}+Pe\frac{\partial}{\partial x}(\phi)-\frac{\partial}{\partial x}\left(\frac{\partial\phi}{\partial x}\right)=0,\quad 0\le x \tag{7.25}$$

where

$$Pe=\frac{v_x L}{\kappa} \tag{7.26}$$

is the Peclet number and L is a convenient length scale. A problem of interest is the diffusion-advection of a step imposed and maintained at the boundary $x = 0$. The appropriate boundary and initial conditions are

$$\phi(x=0,t>0)=\phi_1\ ,\quad \lim_{x\to\infty}\phi=0 \tag{7.27}$$

$$\phi(x>0,t=0)=0 \tag{7.28}$$

The solution is

$$\phi=\frac{\phi_2}{2}\left[erfc\left(\frac{x-Pe\,t}{2\sqrt{t}}\right)+e^{Pe\,x}\,erfc\left(\frac{x+Pe\,t}{2\sqrt{t}}\right)\right] \tag{7.29}$$

where the complimentary error function $erfc(z)=1-erf(z)$ (note $erfc(-z)=1+erf(z)$). Example plots of (7.29), for selected values of the Peclet number, at time $t = 5$, are given by the solid lines in Fig. 7.7.

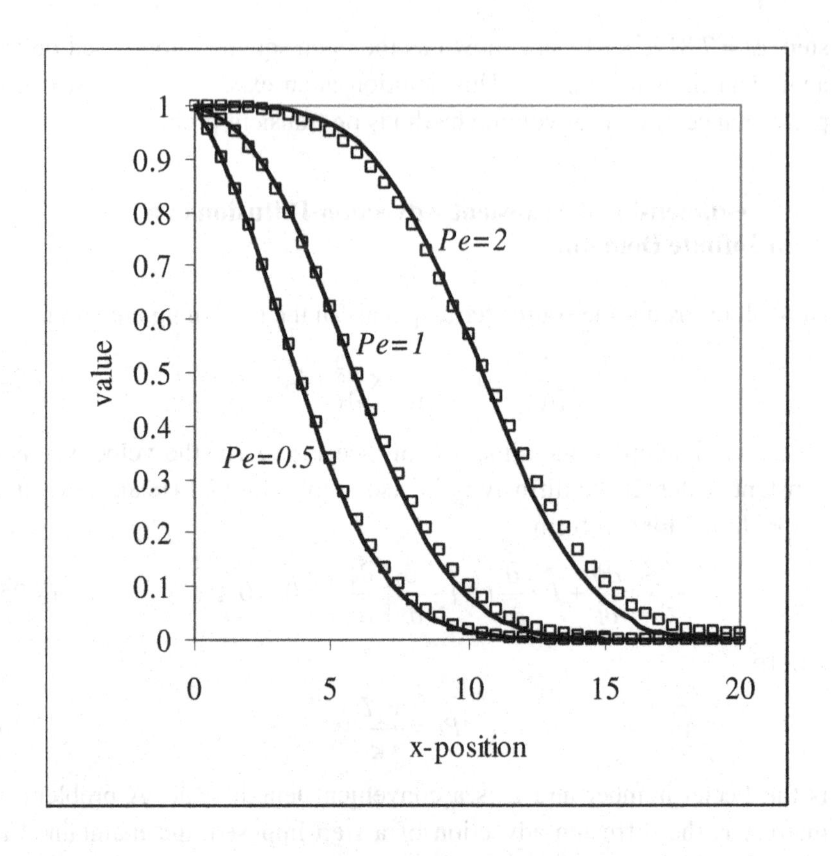

Fig. 7.7 Analytical (solid lines) and CVFEM (open squares) solutions of transient advection-diffusion problem (7.24)

On imposing the boundary conditions

$$\phi(x = 0, -b \le y \le b, t > 0) = \phi_1, \qquad \lim_{x \to \infty} \phi = 0 \tag{7.30a}$$

$$\left.\frac{\partial \phi}{\partial y}\right|_{y=\pm b,\, 0 \le x < \infty} = 0 \tag{7.30b}$$

the initial condition

$$\phi(x > 0, -b \le y \le b, t = 0) = \phi_2 \tag{7.31}$$

and setting the velocities to $v_x = Pe, v_y = 0$ the solution (7.29) is also applicable to the two-dimensional problem

$$\frac{\partial \phi}{\partial t} + \frac{\partial (v_x \phi)}{\partial x} + \frac{\partial (v_y \phi)}{\partial y} - \frac{\partial}{\partial x}\left(\kappa \frac{\partial \phi}{\partial x}\right) - \frac{\partial}{\partial y}\left(\kappa \frac{\partial \phi}{\partial y}\right) = 0, \qquad (7.32)$$

where $0 \le x \le \infty, \quad 0 \le y \le b$.The CVFEM discretization for this problem is also given by (7.23). In setting up this equation the diffusive contributions to the coefficients are calculated from (5.27) and the advective contributions from (5.37). The CVFEM solution has been implemented on the mesh shown in Fig 7.6, where the long length L = 20 is chosen to mimic the far field $x \to \infty$ condition. The settings of the boundary terms are as in the previous problem, i.e., (i) the upper(y = b= 4), the lower(y = 0) and right (far field) boundaries are insulated ($B_C = 0, B_B = 0$) and (ii) the left hand boundary (x = 0) has the fixed value 1 ($B_C = 10^{16}, B_B = 10^{16}$). The CVFEM solutions (profiles along y = 0 for selected Peclet numbers at time t = 5) obtained from the explicit time marching solution of (7.23) with time step $\Delta t = 0.0125$ are shown as the open squares superposed on the analytical lines in Fig. 7.7. At low Peclet numbers (0.5 and 1) the agreement between the CVFEM and the analytical solutions is excellent, indicating the ability of the CVFEM to handle advection transport. As the Peclet number increases, however, we notice a "smearing" in the CVFEM solution. This is due to the "numerical dispersion" upwind feature noted in Chapter 5. The smearing can be mitigated by using a finer mesh or implementing a higher order upwind treatment, see Woodfield, Suzuki and Nakabe (2004).

7.6 Steady State Diffusion in an Annulus

7.6.1 Constant diffusivity

In cylindrical coordinates the governing equation, derived from (7.6), is

$$\frac{d}{dr}\left(r\frac{d\phi}{dr}\right) = 0, \quad R_{in} \le r \le R_{out} \qquad (7.33)$$

where $r = R_{in}$ and $r = R_{out}$ are the inner and out radii of the annulus domain in Fig. 7.2c. The boundary conditions are

$$\phi(r = R_{in}) = \phi_{in} \quad \text{and} \quad \phi(r = R_{out}) = \phi_{out} \tag{7.34}$$

The analytical solution is

$$\phi = \phi_{in} + (\phi_{out} - \phi_{in}) \frac{ln(r) - ln(R_{in})}{ln(R_{out}) - ln(R_{in})} \tag{7.35}$$

A particular solution, when $\phi_{in} = 1, \phi_{out} = 0, R_{in} = 1, R_{out} = 2$, is shown by the dashed line in Fig. 7.8.

In two-dimensional Cartesian coordinates (considering the ¼ domain on the right of Fig. 7.2c the problem statement is

$$\frac{\partial^2 \phi}{\partial x^2} + \frac{\partial^2 \phi}{\partial y^2} = 0, \quad x \geq 0,\ y \geq 0,\ R_{in} \leq \sqrt{x^2 + y^2} \leq R_{out} \tag{7.36}$$

with

$$\phi = \phi_{in},\ on\ \sqrt{x^2 + y^2} = 1,\ \phi = \phi_{out},\ on\ \sqrt{x^2 + y^2} = 2 \tag{7.37}$$

$$\frac{\partial \phi}{\partial y} = 0, \quad on \quad y = 0, \qquad \frac{\partial \phi}{\partial x} = 0, \quad on \quad x = 0 \tag{7.38}$$

The CVFEM discretization for this problem, implemented on the mesh shown in Fig. 7.9, is given by equation (7.13). The coefficients are generated from (5.27). The boundary terms are set such that (i) for nodes on the inner radius (r = 1) the fixed value 1 is imposed by setting $B_C = 10^{16}, B_B = 10^{16}$ (ii) for nodes on the outer radius (r = 2) the fixed value 0 is imposed on setting $B_C = 10^{16}, B_B = 2 \times 10^{16}$ (iii) all other boundaries are insulated, i.e., $B_C = 0, B_B = 0$. The CVFEM solution is obtained by a point wise iteration of (7.13). This two-dimensional solution $\phi(x_i, y_i)$ can be compared to the one-dimensional analytical axisymmetric solution (7.35) on setting

$$r = r_i = \sqrt{x_i^2 + y_i^2}\ .$$

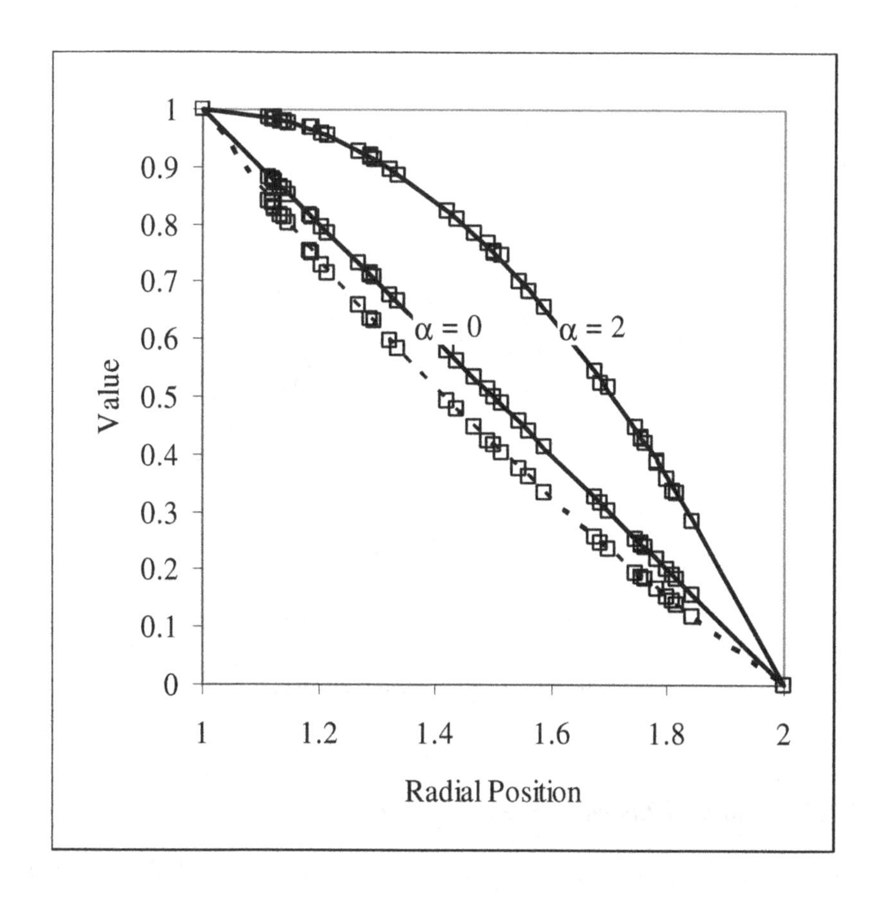

Fig. 7.8 Steady state diffusion solution in annulus, lines axisymmetric solution, symbols CVFEM Cartesian solution

a procedure that will be used thought out the treatment of annulus problems. The CVFEM solution obtained in this way is shown as open squares on Fig. 7.8. Note, essentially the only difference between the CVFEM operation of this solution and the constant diffusion one-dimensional solution discussed in §7.3 is the domain geometry and mesh, i.e., the mesh in Fig. 7.4 vs. the mesh in Fig. 7.9.

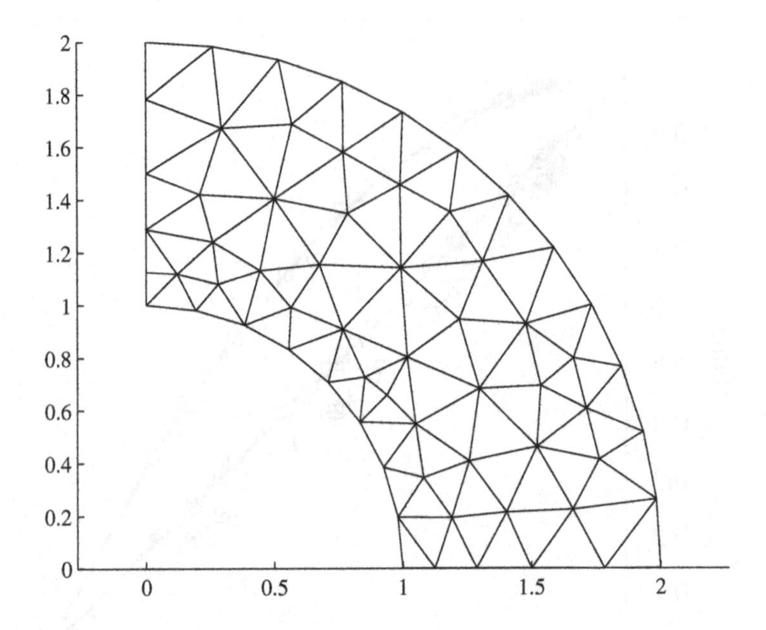

Fig. 7.9 Mesh for annulus problem

7.6.2 Variable diffusivity and source term

The previous problem can be extend by assuming a radial dependent diffusivity and source

$$\kappa = \tfrac{1}{r}, \quad Q = \tfrac{\alpha}{r}, \ \alpha = \text{ constant}$$

In cylindrical coordinates the domain geometry is the left had side of Fig. 7.2c and the governing equation is

$$\frac{d^2\phi}{dr^2} + \alpha = 0, \quad R_{in} \le r \le R_{out} \tag{7.39}$$

With the boundary conditions

$$\phi(r = R_{in}) = \phi_{in} \quad \text{and} \quad \phi(r = R_{out}) = \phi_{out} \tag{7.40}$$

When $\phi_{in} = 1, \phi_{out} = 0, R_{in} = 1, R_{out} = 2$ the analytical solution is

$$\phi = -\frac{\alpha}{2}r^2 - (1-\tfrac{3}{2}\alpha)r + 2 - \alpha \tag{7.41}$$

This solution (with $\alpha = 0$ and 2) is plotted as the solid lines in Fig. 7.8.

In Cartesian coordinates (considering the ¼ domain on the right of Fig. 7.2c) the governing equation is

$$\frac{\partial}{\partial x}\left(\kappa\frac{\partial\phi}{\partial x}\right) + \frac{\partial}{\partial y}\left(\kappa\frac{\partial\phi}{\partial y}\right) + Q = 0,\ x \geq 0,\ y \geq 0,\ 1 \leq \sqrt{x^2+y^2} \leq 2 \tag{7.42}$$

with

$$\phi = 1,\quad on\ \sqrt{x^2+y^2} = 1,\quad \phi = 0,\quad on\ \sqrt{x^2+y^2} = 2 \tag{7.43a}$$

$$\frac{\partial\phi}{\partial y} = 0,\quad on\quad y = 0\quad \frac{\partial\phi}{\partial x} = 0,\quad on\quad x = 0 \tag{7.43b}$$

and

$$\kappa = \frac{1}{\sqrt{x^2+y^2}},\quad Q = \frac{\alpha}{\sqrt{x^2+y^2}} \tag{7.44}$$

This problem, testing the ability of the CVFEM to deal with spatial variability in the diffusivity and source terms, is solved using the full steady sate CVFEM discrete form in (7.3). In solving equations (7.42)-(7.44) this discrete equation is implemented on the mesh shown in Fig. 7.9. The coefficients are calculated using (5.27)—taking care to account for the variable conductivity. The boundary terms are identical to those used in the previous problem in §7.6.1. The source terms in (7.3) are calculated as

$$Q_{C_i} = 0,\quad Q_{B_i} = V_i\frac{\alpha}{\sqrt{x_i^2+y_i^2}} \tag{7.45}$$

Where V_i is the volume of the control volume around node i. The CVFEM solution is obtained by a point wise iteration of (7.3). The CVFEM solution for the cases $\alpha = 0$ and $\alpha = 2$ are shown as open squares in Fig. 7.8.

7.7 Steady State Advection Diffusion in an Annulus

7.7.1 Constant diffusivity

In the case of steady advection-diffusion in an annulus, by continuity the radial velocity scales as $v \sim \frac{1}{r}$. If the scaling constant is unity and the diffusion is the constant $\kappa = 1$, the governing equation, derived from (7.6), is

$$\frac{\partial}{\partial r}(\phi) = \frac{\partial}{\partial r}\left(r\frac{\partial\phi}{\partial r}\right), \quad R_{in} \le r \le R_{out} \tag{7.46}$$

where $r = R_{in}$ and $r = R_{out}$ are the inner and out radii of the annulus domain in Fig 7.2c. The boundary conditions are those of the previous problem

$$\phi(r = R_{in}) = \phi_{in1} \quad \text{and} \quad \phi(r = R_{out}) = \phi_{out} \tag{7.47}$$

The analytical solution is

$$\phi = (\phi_{out} - \phi_{in})r + (2\phi_{in} - \phi_{out}) \tag{7.48}$$

This solution for the case $\phi_{in} = 1, \phi_{out} = 0,\ R_{in} = 1,\ R_{out} = 2$ is plotted as the solid line in Fig. 7.10.

In two-dimensional Cartesian coordinates (considering the ¼ domain on the right of Fig. 7.2c) the problem statement is

$$\frac{\partial v_x\phi}{\partial x} + \frac{\partial v_y\phi}{\partial y} - \frac{\partial}{\partial x}\left(\frac{\partial\phi}{\partial x}\right) - \frac{\partial}{\partial y}\left(\frac{\partial\phi}{\partial y}\right) = 0, \tag{7.49}$$

in

$$x \ge 0,\ y \ge 0,\ R_{in} \le \sqrt{x^2 + y^2} \le R_{out} \tag{7.50}$$

with

$$\phi = \phi_{in},\ on\ \sqrt{x^2 + y^2} = 1,\ \phi = \phi_{out},\ on\ \sqrt{x^2 + y^2} = 2 \tag{7.51}$$

$$\frac{\partial\phi}{\partial y} = 0,\ \ on\ \ y = 0,\ \ \frac{\partial\phi}{\partial x} = 0,\ \ on\ \ x = 0 \tag{7.52}$$

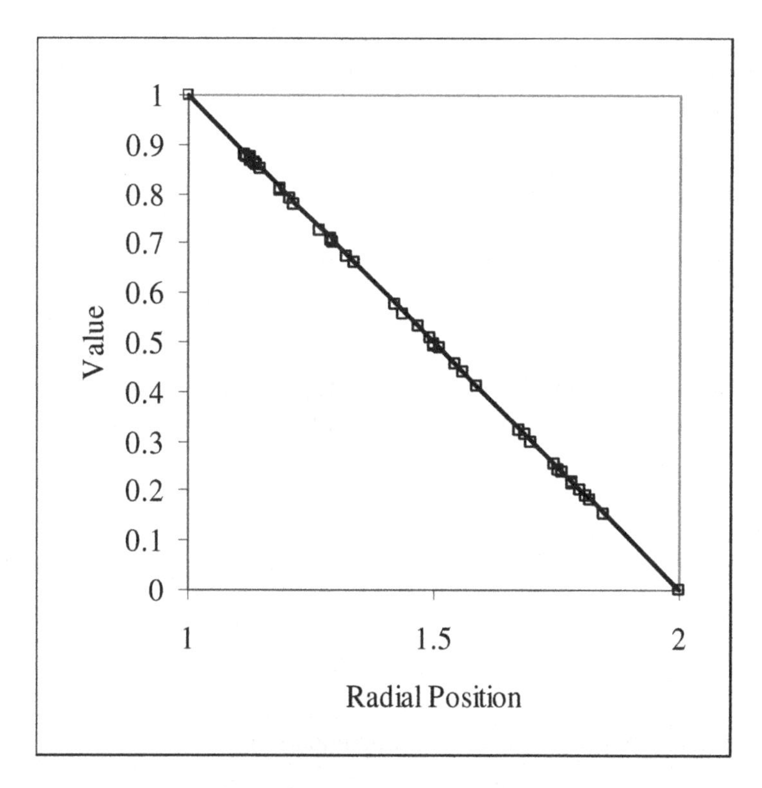

Fig. 7.10 Advection-diffusion in an annulus (constant diffusivity). Solid line axisymmetric analytical solution (7.48). Symbols CVFEM solution of (7.49)-(7.53)

and

$$v_x = \frac{cos(\theta)}{\sqrt{x^2 + y^2}}, \quad v_y = \frac{sin(\theta)}{\sqrt{x^2 + y^2}}, \quad \theta = tan^{-1}(y/x) \qquad (7.53)$$

This problem can be solved using the CVFEM discrete equation in (7.13), the diffusive contributions to the coefficients calculated from (5.27) and the advective contributions from (5.37). This solution can be implemented on the mesh shown in Fig 7.9. For the example problem in Fig. 7.10, the boundary terms are set such that (i) for nodes on the inner radius (r = 1) the fixed value 1 is imposed by setting $B_C = 10^{16}, B_B = 10^{16}$ (ii) for nodes on the outer radius (r = 2) the fixed value 0 is imposed on setting $B_C = 10^{16}, B_B = 2 \times 10^{16}$ (iii) all other boundaries are insulated, i.e., $B_C = 0, B_B = 0$. The CVFEM solution is obtained by a point wise iteration of (7.13). and compared with the analytical solution in Fig. 7.9.

7.7.2 Variable diffusivity

In the previous problem if the diffusivity is given by

$$\kappa = \frac{1}{r},$$

a condition consistent with a dispersion dominated regime (see Hsieh (1986)) the governing equation, in cylindrical coordinates becomes

$$\frac{\partial \phi}{\partial r} = \frac{\partial^2 \phi}{\partial r^2}, \quad R_{in} \le r \le R_{out} \tag{7.54}$$

with boundary conditions identical to (7.47). The analytical solution, when $\phi_{in} = 1, \phi_{out} = 0, R_{in} = 1, R_{out} = 2$, is

$$\phi = \frac{e^r - e^2}{e - e^2} \tag{7.55}$$

In Cartesian coordinates the governing equation is

$$\frac{\partial v_x \phi}{\partial x} + \frac{\partial v_y \phi}{\partial y} - \frac{\partial}{\partial x}\left(\kappa \frac{\partial \phi}{\partial x}\right) - \frac{\partial}{\partial y}\left(\kappa \frac{\partial \phi}{\partial y}\right) = 0, \tag{7.56}$$

Subject to definitions and conditions in (7.50)-(7.53) along with the additional definition

$$\kappa = \frac{1}{\sqrt{x^2 + y^2}}, \tag{7.57}$$

The setting of the variable diffusivity is the only change that needs to be made to the CVFEM code used to solve the previous problem. The comparison of this two-dimensional solution (open symbols) is compared with the analytical solution of (7.55) in Fig. 7.11. The very slight under prediction in the CVFEM solution is due to the relatively coarse mesh used (see Fig. 7.8)). If a finer mesh is used the performance of the CVFEM improves.

Note in Appendix A and B codes to generate a mesh and a CVFEM solution for this problem are provided.

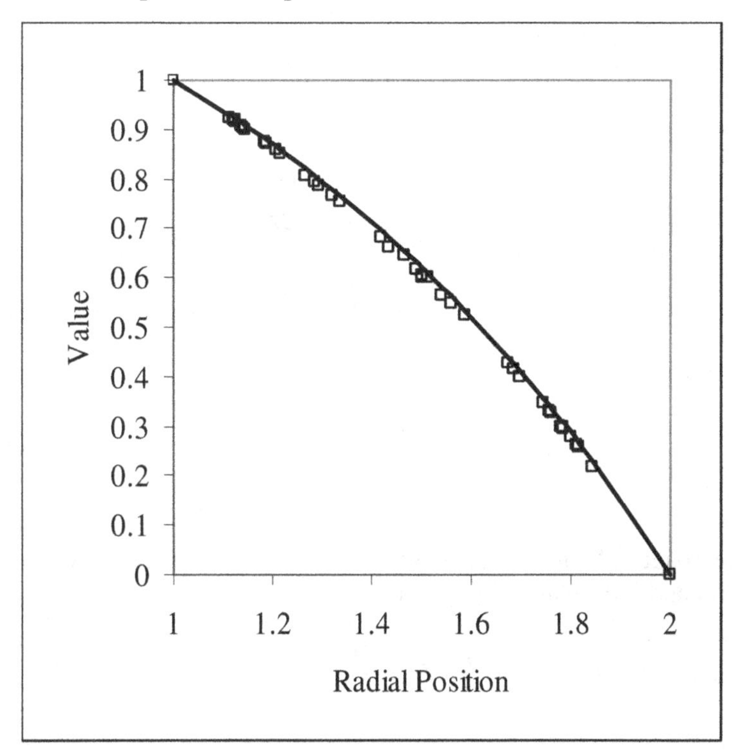

Fig. 7.11 Advection-diffusion in an annulus (variable diffusivity). Solid line axisymmetric analytical solution (7.55). Symbols, CVFEM solution

7.8 Transient Diffusion from a Line Source

7.8.1 Problem

This problem involves a line source of strength Q_ℓ (quantity/time-unit depth) placed at the origin. As time increases there is a radial flow of quantity out form the source. The domain of the problem is shown in the left of Fig. (7.2a). With a constant diffusivity $\kappa = 1$, the governing equation is the transient cylindrical diffusion equation

$$r\frac{\partial\phi}{\partial t}=\frac{\partial}{\partial r}\left(r\frac{\partial\phi}{\partial r}\right), \quad r>0 \tag{7.58}$$

with the initial condition

$$\phi(r,0)=0 \tag{7.59}$$

and the boundary conditions

$$\lim_{r\to 0}\left[r\frac{\partial\phi}{\partial r}\right]=-\frac{Q_\ell}{2\pi} \tag{7.60a}$$

$$\lim_{r\to\infty}\phi=0 \tag{7.60b}$$

The analytical solution is

$$\phi=\frac{Q_\ell}{4\pi}E_1\left(\frac{r^2}{4t}\right) \tag{7.61}$$

where (page 228 in Abramowitz and Stegun (1970)) the exponential integral and it derivative are given by

$$E_1(z)=\int_z^\infty\frac{e^{-\alpha}}{\alpha}d\alpha \tag{7.62}$$

and

$$\frac{dE_1}{dz}=-\frac{e^{-z}}{z} \tag{7.63}$$

Values of the exponential integral can be estimated, from the following rational approximation given in Abramowitz and Stegun (1970)

$$\begin{aligned}
&0\le z\le 1\\
&E_1(z)+ln\,z=a_0+a_1z+a_2z^2+a_3z^3+a_4z^4+a_5z^5\\
&a_0=-.57721566, a_1= .99999193\ \ a_2=-.24991055,\\
&a_3= .05519968, a_4=-.00976004, a_5= .00107857
\end{aligned} \tag{7.64}$$

and

$$
\begin{aligned}
&1 \le z \\
&ze^{z}E_1(z) = \frac{z^2 + a_1 z + a_2}{z^2 + b_1 z + b_2} \\
&a_1 = 2.334733, b_1 = 3.330657 \\
&a_2 = .250621, b_2 = 1.681534
\end{aligned}
\tag{7.65}
$$

where the absolute errors are bounded by 2×10^{-7} and 5×10^{-5} respectively.

In Cartesian coordinates the governing equation is

$$\frac{\partial\phi}{\partial t} - \frac{\partial^2\phi}{\partial x^2} - \frac{\partial^2\phi}{\partial y^2} - Q = 0, \quad 0 \le x \le L,\ 0 \le y \le L \tag{7.66}$$

The domain is the ¼ square region shown on the right of Fig. 7.2a. Insulated conditions are imposed on all the boundaries, and L is chosen large enough such that, over the solution time, the far field condition (7.60b) is met. The line source problem is recovered from (7.66) on applying the initial condition $\phi = 0$ and defining the source to be

$$Q = \tfrac{1}{4} Q_l \delta(x-0, y-0) \tag{7.67}$$

where δ is the Dirac delta function, see (5.45).

7.8.2 Unstructured mesh solutions

The CVFEM solution is based on the discrete equation (7.4), i.e.,

$$(V_i + B_{C_i})\phi_i^{new} = V_i\phi_i + \Delta t\left(\sum_{j=1}^{n_i} a_{i,j}\phi_{S_{i,j}} - a_i\phi_i + Q_{B_i} - Q_{C_i}\phi_i\right) + B_{B_i}$$

Where V_i is the volume of the ith control volume and the coefficients are calculated with (5.27). The default, insulated setting, $B_C = 0, B_B = 0$, is made on all the boundaries. The source terms are set as

$$
\begin{aligned}
&Q_{C_i} = 0 \\
&Q_{B_i} = \begin{cases} \dfrac{Q_l}{4}, \text{if } i = 1 \\ 0, \ \text{otherwise} \end{cases}
\end{aligned}
\tag{7.68}
$$

where $i=1$ is the node at the origin $x=0, y=0$. The solution, the explicit solution of (7.4) using a time step $\Delta t = 0.0125$, is implemented on the unstructured mesh shown in Fig. 7.12

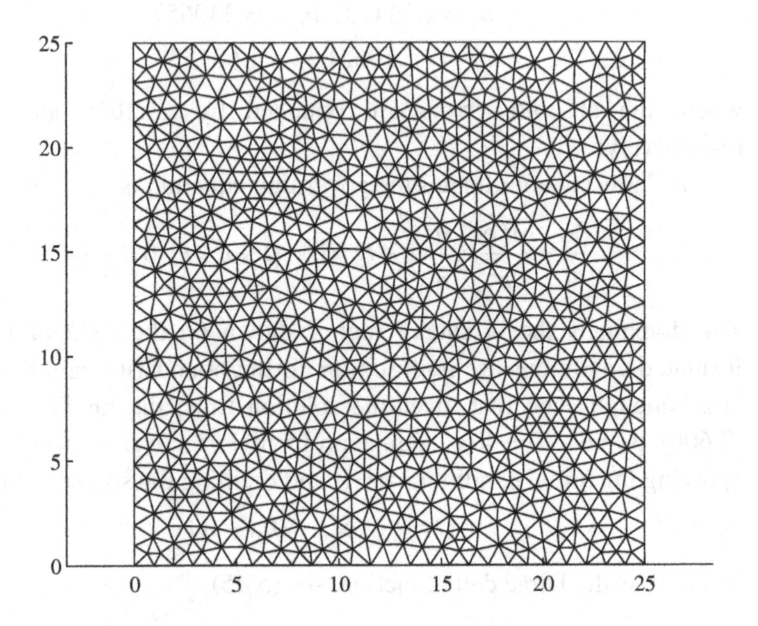

Fig. 7.12 Unstructured mesh for line source problem

The CVFEM results, profiles of ϕ along the x-axis at times t = 12.5, 25. and 50, are compared with the analytical solution (7.61), for the case $Q_\ell = 1$, in Fig. 7.13. Note, to more clearly see the comparison a log scale has been used for the value of ϕ. The agreement is excellent clearly showing the performance power of the CVFEM in solving two-dimensional transient field problems.

7.8.3 Structured mesh solutions

The current test problem can also be solved using structured approaches, in particular the CVFEM based on a structured mesh and the control volume finite difference method (Chapter 6) based on a structured grid.

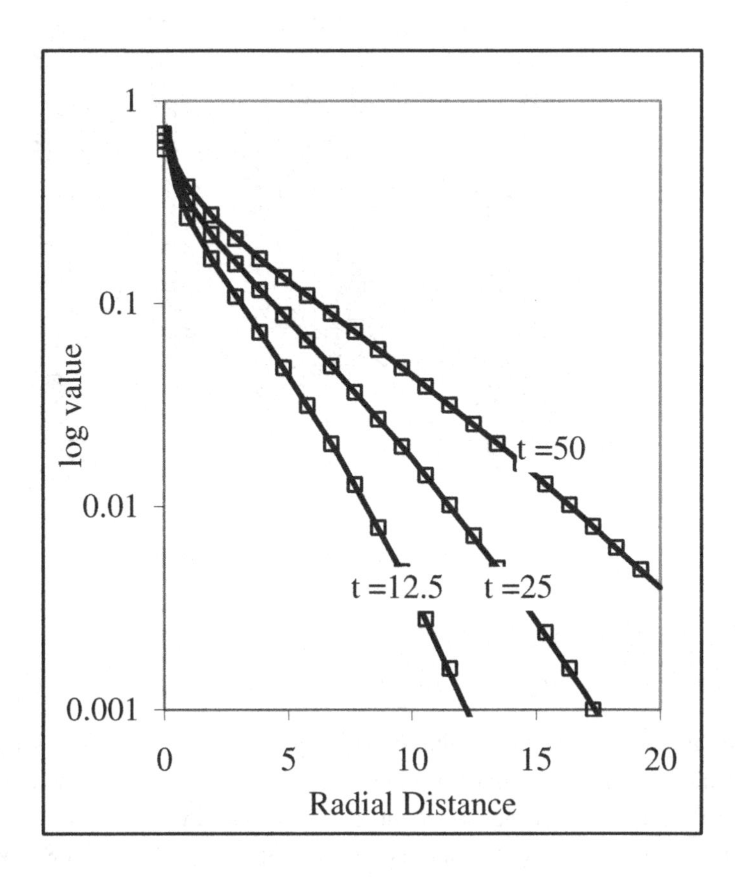

Fig. 7.13 Results for line source problem when $Q_\ell = 1$. Solid line analytical solution (7.61), open symbols CVFEM solution implemented on mesh in Fig. 7.12

CVFEM on a structured grid: The CVFEM solution above can immediately be implemented on the structured mesh shown in Fig. 7.14. The results, profile of ϕ along the x-axis at time t = 50, are compared with the analytical solution (7.61), for the case $Q_\ell = 1$, in Fig. 7.15.

CVFDM solution: Taking guidance from the work in Chapter 6, the control volume finite difference method (CVFDM) applied to the governing equation (7.66) has the discrete explicit transient form

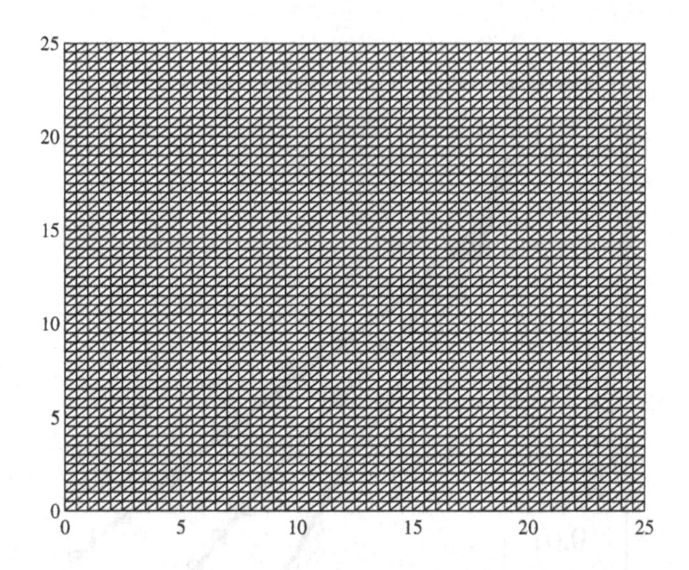

Fig. 7.14 Structured CVFEM mesh

$$\Delta x_{ij}\Delta y_{ij}\phi_{i,j}^{new} = \Delta x_{ij}\Delta y_{ij}\phi_{i,j} + \Delta t\left[a_{n_{ij}}\phi_{i+1,j} + a_{w_{ij}}\phi_{i,j-1} + a_{s_{ij}}\phi_{i-1,j} + a_{e_{ij}}\phi_{i,j+1} - a_{p_{i,j}}\phi_{i,j} + Q_{ij}\right] \quad (7.69)$$

An appropriate control volume grid is given in Fig 7.16. Working on this grid the coefficients in (7.69) are calculated using (6.16). The appropriate coefficients in volumes adjacent to the insulated boundary are set to zero. The source term Q_{ij} is set using the second part of (7.68). With the coefficients and source terms calculated (7.79) can be solved by explicit time marching using a step of $\Delta t = 0.025$. The results (profile along the lowest layer of node points in Fig. 7.15)) obtained in this manner are compared with the analytical solution (7.61) in Fig. 7.15.

7.9 The Recharge Well Problem

A vertical recharge well of radius $r = r_w$ issues a flow Q (volume/time) into a horizontal uniformly thick (b) porous (porosity ε) aquifer. At time $t = 0$ a concentration of ϕ_0 is introduce and held fixed in the well. This concentration is transported into the aquifer by both advection and hydraulic dispersion. In the radial direction the flow velocity is

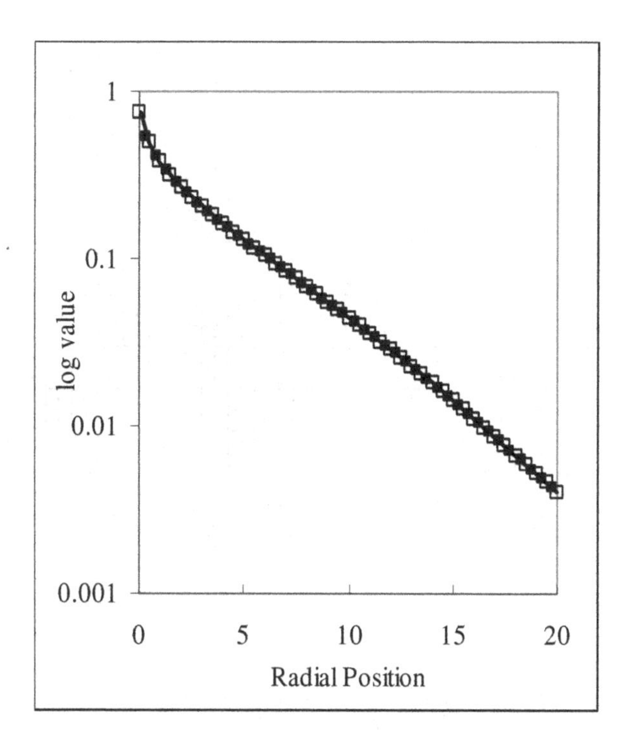

Fig. 7.15 Structured grid Results at $t = 50$. Solid line analytical, Open symbols CVFEM using mesh in Fig. 7.14, Closed solid symbols CVFDM using grid in Fig 7.16

$$v_r(r) = \frac{A}{r}, \quad A = \frac{Q}{2\pi b\varepsilon} \tag{7.70}$$

and the dispersion is modeled by

$$\kappa = \alpha \frac{A}{r} \tag{7.71}$$

where α is the longitudinal dipersivity. Working from Hsieh (1986), on substitution in to (7.6), the governing equation, in polar cylindrical coordinates, is

$$\frac{\partial\phi}{\partial t} + \frac{A}{r}\frac{\partial}{\partial r}(\phi) = \alpha \frac{A}{r}\frac{\partial^2\phi}{\partial r^2}, \quad r > r_w \tag{7.72}$$

with the following initial and boundary conditions

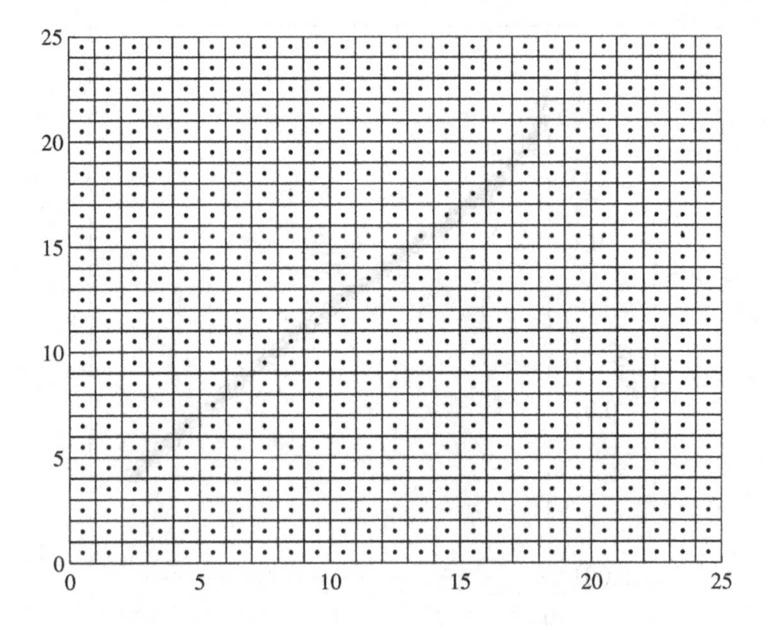

Fig. 7.16 Control volume finite difference grid

$$\phi(r>0,0)=0, \quad \phi(r_w,t>0)=\phi_0, \quad \phi(r\to\infty,t>0)=0 \quad (7.73)$$

The analytical solution can be obtained via a Lapalce transform. This solution is too detailed to presents here. Fortunately, however, Hsieh (1986) has tabulated numerical values of the solution for various choices of the problem parameters; a selection of these tabulations is reproduced in Table 7.1.

In two-dimensional Cartesian coordinates the governing equation is given by (7.32)

$$\frac{\partial\phi}{\partial t}+\frac{\partial(v_x\phi)}{\partial x}+\frac{\partial(v_y\phi)}{\partial y}-\frac{\partial}{\partial x}\left(\kappa\frac{\partial\phi}{\partial x}\right)-\frac{\partial}{\partial y}\left(\kappa\frac{\partial\phi}{\partial y}\right)=0, \quad x^2+y^2\geq r_w$$

The domain (a ¼ section) and an unstructured mesh for this problem when $r_w = 10$ is shown in Fig. 7.17. In this domain the initial condition is $\phi(x,y)=0$, all the boundaries are set as no-flow boundaries apart from the curved well surface where

Table 7.1 Tabulated analytical results for re-charge well problem for conditions, $r_w = \phi_0 = \alpha = A = 1$ (after Hsieh (1986))

$\tau = 50$		$\tau = 100$		$\tau = 500$	
ρ	ϕ	ρ	ϕ	ρ	ϕ
11	0.964	11	0.993	22	0.99
12	0.892	13	0.949	26	0.936
13	0.775	14	0.9	28	0.867
13.5	0.701	15	0.826	30	0.757
14	0.617	16	0.724	31	0.686
14.5	0.529	17	0.6	32	0.607
15	0.439	18	0.463	33	0.523
15.5	0.353	19	0.329	34	0.436
16	0.273	20	0.213	35	0.352
16.5	0.203	21	0.124	36	0.274
17	0.145	22	0.065	38	0.148
18	0.064	23	0.03	40	0.067
19	0.023	25	0.004	44	0.008

$$\phi(x^2 + y^2 = r_w) = 1 \tag{7.74}$$

The extend of the domain is chosen to be large enough that the imposition of no-flow boundaries, for the time duration of the simulation, is a reasonable mimic for the far field condition $\phi \to 0$. The appropriate values for the components of the velocity field and diffusivity are given in equations (7.53) and (7.57).

The CVFEM solution is based on the discrete equation (7.4), i.e., neglecting the source terms

$$(V_i + B_{C_i})\phi_i^{new} = V_i\phi_i + \Delta t\left(\sum_{j=1}^{n_i} a_{i,j}\phi_{S_{i,j}} - a_i\phi_i\right) + B_{B_i} \tag{7.75}$$

Where V_i is the volume of the ith control volume and the coefficients are calculated with (5.27) and (5.37). All, apart from the well boundary are insulated $B_C = 0, B_B = 0$, on the well boundary $B_C = 10^{16}, B_B = 10^{16}$. Fig. 7.18 compares predictions, profiles along the x-axis at selected times, using (7.75) with the analytical solutions tabulated in Table 7.1.

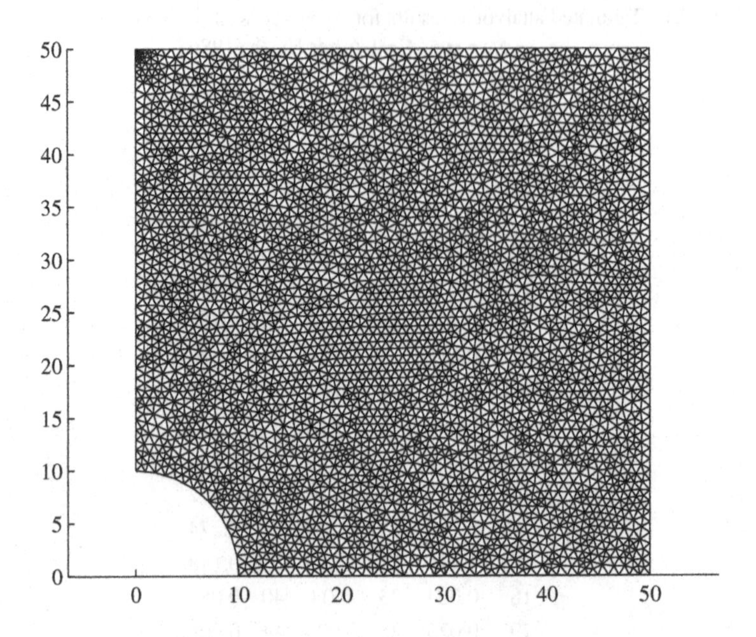

Fig. 7.17 Domain and mesh for recharge well problem

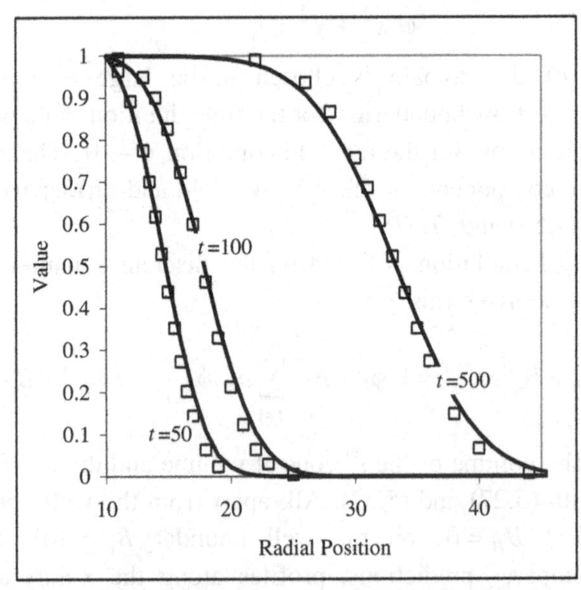

Fig. 7.18 Profiles of ϕ along the x-axis at selected time for the recharge well problem. Symbols the analytical solution Table 7.1, lines the numerical predictions from (7.75)

The comparison in Fig. 7.18 clearly shows the ability of the CVFEM to deal with a general transient two-dimensional advection –diffusion problem. The slight disagreement is attributed to artificial dispersion (numerical dissipation) which is a feature of the grid size.

Chapter 8

A Plane Stress CVFEM Solution

The control volume finite element method (CVFEM) is applied to the solution of a plane stress problem.

8.1 Introduction

The task in this chapter is outline how the CVFEM can be applied in the solution of a linear elastic problem. The problem chosen is the stress concentration due to a hole in an infinite plate subjected to a uniaxial load. The problem will be cast and solved as a plane stress problem. Recall however, by the result in §2.2.6, a plane stress solution can be converted to a plane strain solution through an appropriate re-definition of the Young's modulus, Poisson's ratio, and thermal expansion coefficient.

8.2 The Stress Concentration Problem

The problem domain is shown in Figure 8.1. This domain consists of a ¼ of finite plate of dimensions 200×200 mm with a center hole of radius $a = 10mm$. On the right side of the domain a uniaxial (x-direction) stress $\sigma_0 = 1\, N/mm^2$ is applied. The top side of the domain and the hole is unconstrained. The left boundary is constrained in the x –direction and the bottom boundary in the y-direction. Following Kikuchi (1986), The Young's modulus is $E = 2.1\times10^5\, N/mm^2$, the Poisson's ratio is $\nu = 0.29$, and the plate has a nominal thickness t = 10mm. There is a classical solution for this problem in cylindrical polar coordinates due to Timoshenko and Goodier (see page 231 in Kikuchi (1986)).

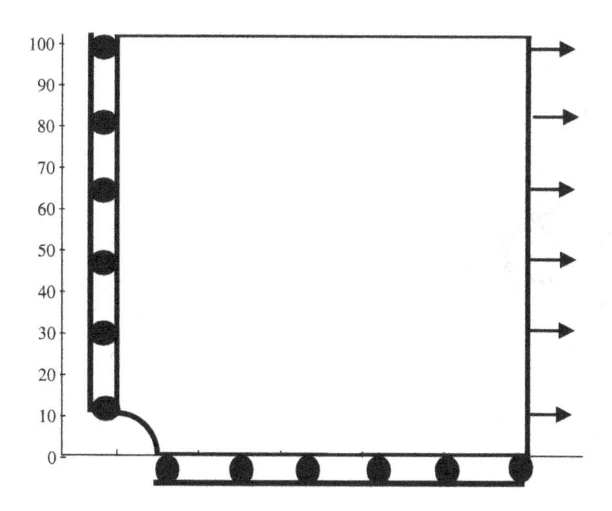

Fig. 8.1 Domain for stress concentration problem

This solution can be used to analytically determine the stress concentration $\sigma_c = \sigma_{xx} / \sigma_0$ along the y-axis

$$\sigma_c = \frac{1}{2}\left\{\left[1 + \left(\frac{a}{y}\right)^2\right] + \left[1 + 3\left(\frac{a}{y}\right)^4\right]\right\}, \quad x = 0, y \geq a \tag{8.1}$$

8.3 CVFEM Displacement Solution

In terms of the displacements (u_x, u_y) the Cartesian statement of this problem, derived from (2.29) can be written as

$$\oint_S \kappa_{xx} \frac{\partial u_x}{\partial x} n_x + \kappa_{xy} \frac{\partial u_x}{\partial y} n_y + \nu \kappa_{xx} \frac{\partial u_y}{\partial y} n_x + \kappa_{xy} \frac{\partial u_y}{\partial x} n_y \, dS = 0 \tag{8.2a}$$

$$\oint_S \kappa_{xy} \frac{\partial u_y}{\partial x} n_x + \kappa_{yy} \frac{\partial u_y}{\partial y} n_y + \kappa_{xy} \frac{\partial u_x}{\partial y} n_x + \nu \kappa_{yy} \frac{\partial u_x}{\partial x} n_y \, dS = 0 \qquad (8.2b)$$

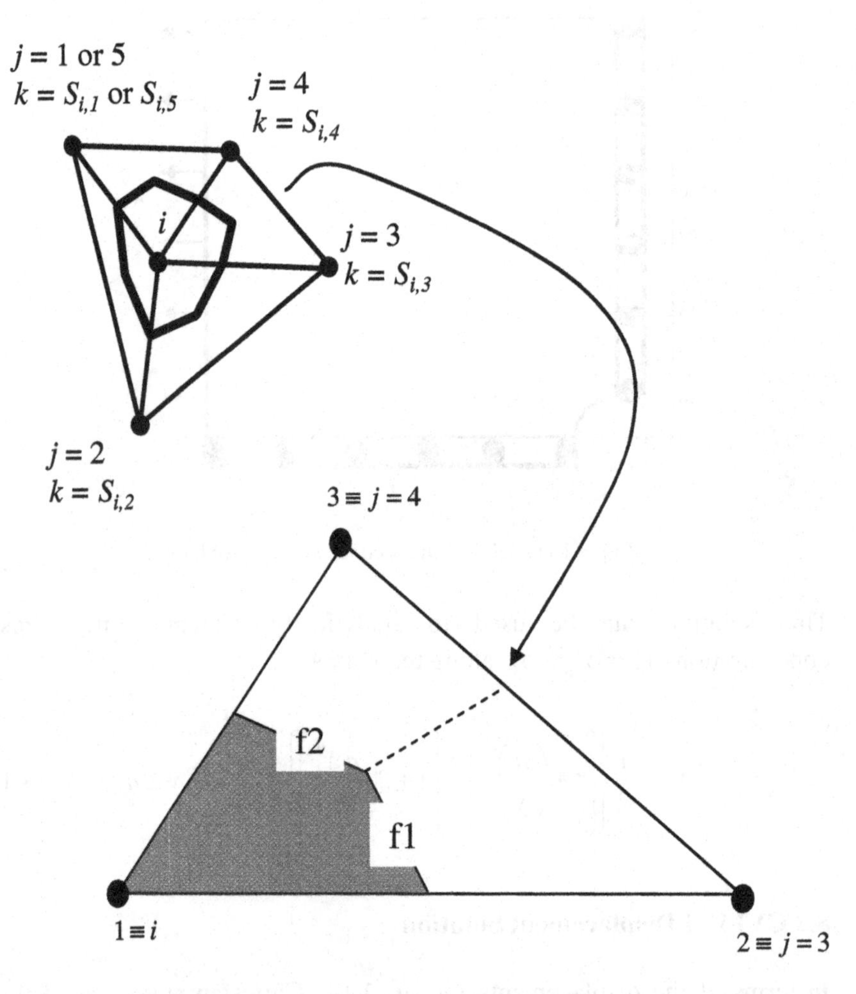

Fig. 8.2 Arrangement of control volumes and elements

where

$$\kappa_{xx} = \kappa_{yy} = E/(1-\nu^2), \quad \kappa_{xy} = E/[2(1+\nu)] \qquad (8.3)$$

8.3.1 The CVFEM discrete equations

A CVFEM solution of (8.2) is sought. The first step is to cover the domain with a mesh of triangular elements, associate the domain of integration in (8.2) with control volumes constructed on this grid, see Fig. 8.2, and then use finite element interpolations to approximate the integrations in terms of nodal values in the region of support. The objective is to end up with the following discrete forms of (8.2)

$$[\, a_i^x + B_{C_i}^x \,] u_{x_i} = \sum_{j=1}^{n_i} a_{i,j}^x u_{x S_{i,j}} + \sum_{j=1}^{n_i} b_{i,j}^x u_{y S_{i,j}} - b_i^x u_{y_i} + B_{B_i}^x \quad (8.4a)$$

$$[\, a_i^y + B_{C_i}^y \,] u_{y_i} = \sum_{j=1}^{n_i} a_{i,j}^y u_{y S_{i,j}} + \sum_{j=1}^{n_i} b_{i,j}^y u_{x S_{i,j}} - b_i^y u_{x_i} + B_{B_i}^y \quad (8.4b)$$

where, using the notations introduced previously, n_i is the number of nodes in the support of node i, $S_{i,j}$ is the global index of the jth node in the support of the ith node, the a's and b's are coefficients, and the B's represent contributions from boundaries. Following the template for diffusion coefficients laid out in §5.3 the coefficients for (8.4a), associated with the faces $f1$ and $f2$ of the element highlighted in Fig. 8.2, are

$$\begin{aligned}
a_1^x &= -\kappa_{xx} N_{1x}(\Delta\vec{y}_{f1} + \Delta\vec{y}_{f2}) + \kappa_{xy} N_{1y}(\Delta\vec{x}_{f1} + \Delta\vec{x}_{f2}) \\
a_2^x &= \kappa_{xx} N_{2x}(\Delta\vec{y}_{f1} + \Delta\vec{y}_{f2}) - \kappa_{xy} N_{2y}(\Delta\vec{x}_{f1} + \Delta\vec{x}_{f2}) \\
a_3^x &= \kappa_{xx} N_{3x}(\Delta\vec{y}_{f1} + \Delta\vec{y}_{f2}) - \kappa_{xy} N_{3y}(\Delta\vec{x}_{f1} + \Delta\vec{x}_{f2})
\end{aligned} \quad (8.5a)$$

$$\begin{aligned}
b_1^x &= -\nu\kappa_{xx} N_{1y}(\Delta\vec{y}_{f1} + \Delta\vec{y}_{f2}) + \kappa_{xy} N_{1x}(\Delta\vec{x}_{f1} + \Delta\vec{x}_{f2}) \\
b_2^x &= \nu\kappa_{xx} N_{2y}(\Delta\vec{y}_{f1} + \Delta\vec{y}_{f2}) - \kappa_{xy} N_{2c}(\Delta\vec{x}_{f1} + \Delta\vec{x}_{f2}) \\
b_3^x &= \nu\kappa_{xx} N_{3y}(\Delta\vec{y}_{f1} + \Delta\vec{y}_{f2}) - \kappa_{xy} N_{3x}(\Delta\vec{x}_{f1} + \Delta\vec{x}_{f2})
\end{aligned} \quad (8.5b)$$

where , moving counter-clockwise around node i, the signed distances

$$\Delta\vec{x}_{f1} = \frac{x_3}{3} - \frac{x_2}{6} - \frac{x_1}{6}, \; \Delta\vec{x}_{f2} = -\frac{x_2}{3} + \frac{x_3}{6} + \frac{x_1}{6} \quad (8.6a)$$

$$\Delta\vec{y}_{f1} = \frac{y_3}{3} - \frac{y_2}{6} - \frac{y_1}{6}, \quad \Delta\vec{y}_{f2} = -\frac{y_2}{3} + \frac{y_3}{6} + \frac{y_1}{6} \tag{8.6b}$$

the derivatives of the shape functions

$$\begin{aligned}
N_{1x} &= \frac{\partial N_1}{\partial x} = \frac{(y_2 - y_3)}{2V^{ele}}, & N_{1y} &= \frac{\partial N_1}{\partial y} = \frac{(x_3 - x_2)}{2V^{ele}} \\
N_{2x} &= \frac{\partial N_2}{\partial x} = \frac{(y_3 - y_1)}{2V^{ele}}, & N_{2y} &= \frac{\partial N_2}{\partial y} = \frac{(x_1 - x_3)}{2V^{ele}} \\
N_{3x} &= \frac{\partial N_3}{\partial x} = \frac{(y_1 - y_2)}{2V^{ele}}, & N_{3y} &= \frac{\partial N_3}{\partial y} = \frac{(x_2 - x_1)}{2V^{ele}}
\end{aligned} \tag{8.7}$$

and the volume of the element

$$V^{ele} = \frac{(x_2 y_3 - x_3 y_2) + x_1(y_2 - y_3) + y_1(x_3 - x_2)}{2} \tag{8.8}$$

In a similar manner the coefficients for (8.4b) are

$$\begin{aligned}
a_1^y &= -\kappa_{xy} N_{1x}(\Delta\vec{y}_{f1} + \Delta\vec{y}_{f2}) + \kappa_{yy} N_{1y}(\Delta\vec{x}_{f1} + \Delta\vec{x}_{f2}) \\
a_2^y &= \kappa_{xy} N_{2x}(\Delta\vec{y}_{f1} + \Delta\vec{y}_{f2}) - \kappa_{yy} N_{2y}(\Delta\vec{x}_{f1} + \Delta\vec{x}_{f2}) \\
a_3^y &= \kappa_{xy} N_{3x}(\Delta\vec{y}_{f1} + \Delta\vec{y}_{f2}) - \kappa_{yy} N_{3y}(\Delta\vec{x}_{f1} + \Delta\vec{x}_{f2})
\end{aligned} \tag{8.9a}$$

$$\begin{aligned}
b_1^y &= -\kappa_{xy} N_{1y}(\Delta\vec{y}_{f1} + \Delta\vec{y}_{f2}) + \nu\kappa_{yy} N_{1x}(\Delta\vec{x}_{f1} + \Delta\vec{x}_{f2}) \\
b_2^y &= \kappa_{xy} N_{2y}(\Delta\vec{y}_{f1} + \Delta\vec{y}_{f2}) - \nu\kappa_{yy} N_{2c}(\Delta\vec{x}_{f1} + \Delta\vec{x}_{f2}) \\
b_3^y &= \kappa_{xy} N_{3y}(\Delta\vec{y}_{f1} + \Delta\vec{y}_{f2}) - \nu\kappa_{yy} N_{3x}(\Delta\vec{x}_{f1} + \Delta\vec{x}_{f2})
\end{aligned} \tag{8.9b}$$

The values in (8.5) and (8.9) can be used to *update* the ith support coefficients of Fig. 8.2 through

$$\begin{aligned}
&a_i^x = a_i^x + a_1^x, \quad b_i^x = b_i^x + b_1^x, \quad a_i^y = a_i^y + a_1^y, \quad b_i^y = b_i^y + b_1^y \\
&a_{i,3}^x = a_{i,3}^x + a_2^x, \quad b_{i,3}^x = b_{i,3}^x + b_2^x, \quad a_{i,3}^y = a_{i,3}^y + a_2^y, \quad b_{i,3}^y = b_{i,3}^y + b_2^y \\
&a_{i,4}^x = a_{i,4}^x + a_3^x, \quad b_{i,4}^x = b_{i,4}^x + b_3^x, \quad a_{i,4}^y = a_{i,4}^y + a_3^y, \quad b_{i,4}^y = b_{i,4}^y + b_3^y
\end{aligned} \tag{8.10}$$

In this way, by calling at each element in the support of node i, the coefficients in (8.4) can be calculated.

8.3.2 Boundary conditions

The coefficients derived above account for the tractions acting on the *internal faces* of a given control volume. To complete the discrete equations (8.4) it is necessary to add in the applied tractions on the right hand vertical boundary ($x = 100mm$). Fig. 8.3 shows a typical control volume on this boundary

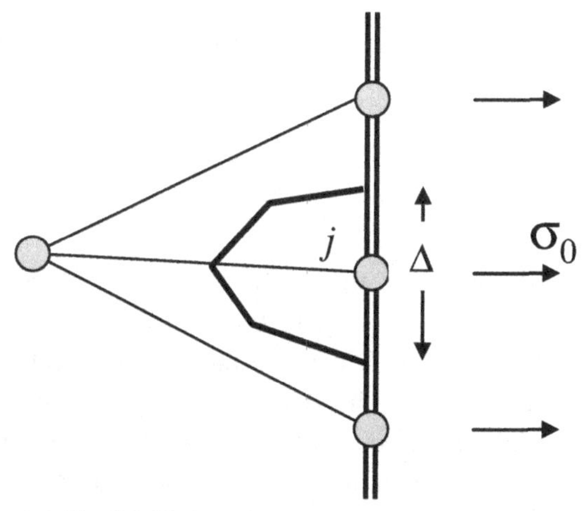

Fig. 8.3 Node and control volume on traction boundary

With reference to this figure the traction boundary is imposed by setting at each node j on this boundary

$$B_B^x = \Delta_j \sigma_0, \quad B_C^x = B_B^y = B_C^y = 0 \tag{8.11}$$

where the lengths Δ_j are calculated from

$$\Delta_j = \begin{cases} (y_{B_2} - y_{B_1})/2 \\ (y_{B_{j+1}} - y_{B_{j-1}})/2, \quad 2 \le j \le n_B - 1 \\ (y_{Bn_B} - y_{Bn_B-1})/2 \end{cases} \tag{8.12}$$

n_B is the number of nodes on the boundary segment where the traction is applied , and B_j is the global node number of the jth node on the boundary.

On each node on the left hand vertical boundary $x = 0$ the zero x-displacement condition is applied by setting

$$B_C^x = 10^{16}, \quad B_B^x = B_B^y = B_C^y = 0 \tag{8.13}$$

On each node the lower horizontal boundary $y = 0$ the zero y-displacement condition is applied by setting

$$B_C^y = 10^{16}, \quad B_B^x = B_B^y = B_C^x = 0 \tag{8.14}$$

On all other boundaries no constraints are imposed, i.e., at each node

$$B_B^x = B_B^y = B_C^x = B_C^y = 0 \tag{8.15}$$

8.3.3 Solution

The work above fully specifies the coefficients in (8.4). Solution, the nodal displacement field (u_x, u_y) can be readily achieved by a point iteration. Figure 8.4 graphically illustrates the displacement solution by showing, in exaggerated form, the deformed shape of the plate after application of the uniaxial stress; this figure also shows the grid used in the calculation. When a larger hole ($a = 40mm$ instead of $a = 10mm$) is used the deformations are much more pronounced, Fig. 8.5. Note with the larger hole the top free surface undergoes a rotation not seen with the smaller hole. This would suggest that with the larger hole the comparison with the analytical stress solution (8.1), requiring an infinite plate, may not be valid. This will be confirmed in the next section where it is shown how to calculate the stress field associated with the solution of (8.4)

8.4 The Stress Solution

In order to compare the solution from (8.4) with the analytical solution in (8.1) it is necessary to calculate the stress filed associated with the displacement fields solved by (8.4).

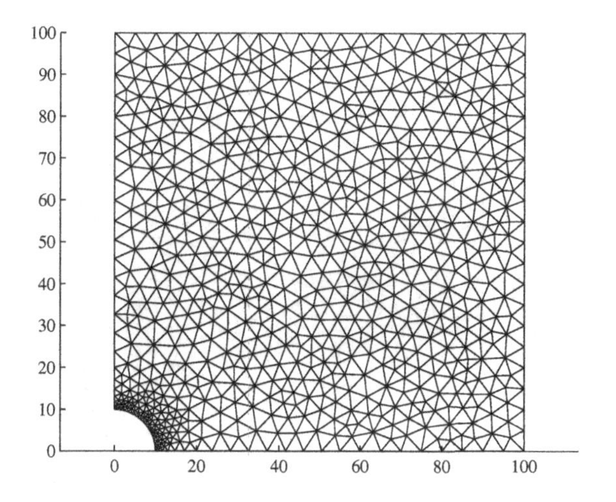

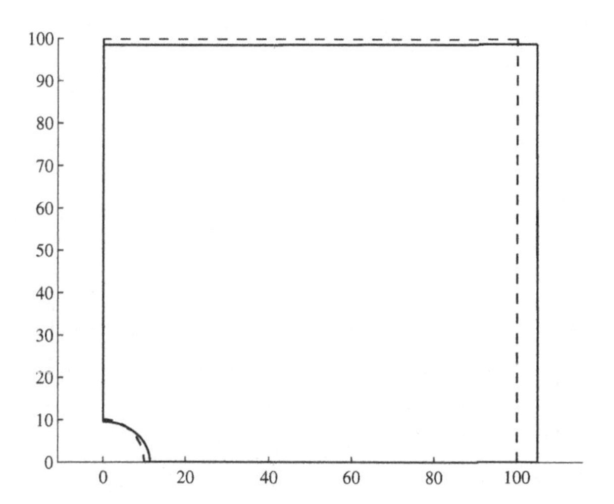

Fig. 8.4 Grid and displacement for *a =10mm*

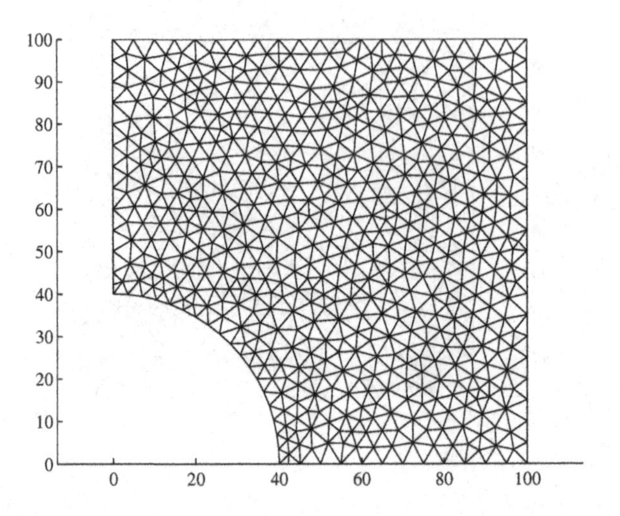

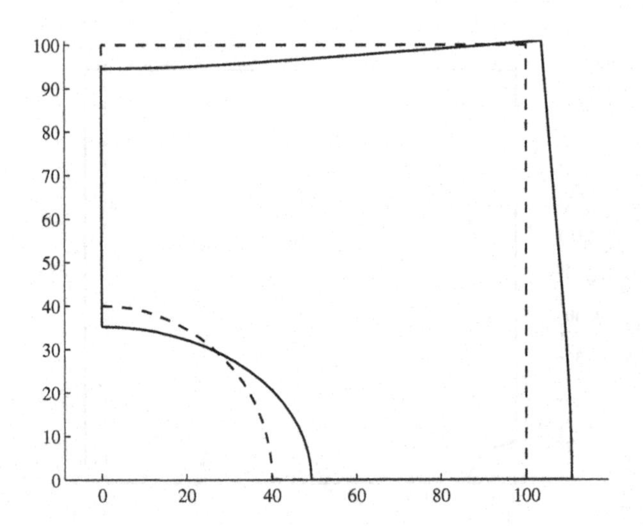

Fig. 8.5 Grid and displacement for $a = 40mm$

8.4.1 Stress in an element

The stress can be determined from the expressions given in (2.32), e.g., in the plane stress case

$$\sigma_{xx} = \frac{E}{(1-\nu^2)}\left[\frac{\partial u_x}{\partial x} + \nu\frac{\partial u_y}{\partial y}\right] \tag{8.16}$$

The CVFEM plane stress solution (8.4) provides the nodal displacement fields $\boldsymbol{u}_x, \boldsymbol{u}_y$. With linear elements the derivatives in (8.16) are constant over a given element, e.g., for the element defined by the nodal vertices labeled 1-2-3, in Fig. 8.6,

$$\left.\frac{\partial u_x}{\partial x}\right|_{ele} = N_{1x}u_{x_1} + N_{2x}u_{x_2} + N_{3x}u_{x_3} \tag{8.17a}$$

$$\left.\frac{\partial u_y}{\partial y}\right|_{ele} = N_{1y}u_{y_1} + N_{2y}u_{y_2} + N_{3y}u_{y_3} \tag{8.17b}$$

where the derivatives of the shape function are constant expressions given by (8.7). Use of (8.17) in (8.16) will readily provide a means of calculating a representative stress for a given element. In many cases, however, we would like to estimate and associate a representative stress with a node.

8.4.2 Estimation of a nodal derivative

In order to calculate a representative stress at a node from the given nodal displacement fields it is necessary to have a way to estimate nodal derivatives of a variable ϕ in terms of its nodal field. With reference to Fig. 8.6 the divergence thereon can be used to write

$$\left.\frac{\partial \phi}{\partial x}\right|_1 V_1 = \int_{CV} \frac{\partial \phi}{\partial x} dA = \oint_{CS} \phi n_x dS \approx \sum_{faces} \phi_{mid} \Delta \bar{y} \tag{8.18}$$

where V_1 is the volume (area) of the control volume around node 1 (shown as the doted line), ϕ_{mid} is an estimate of the field value at the mid

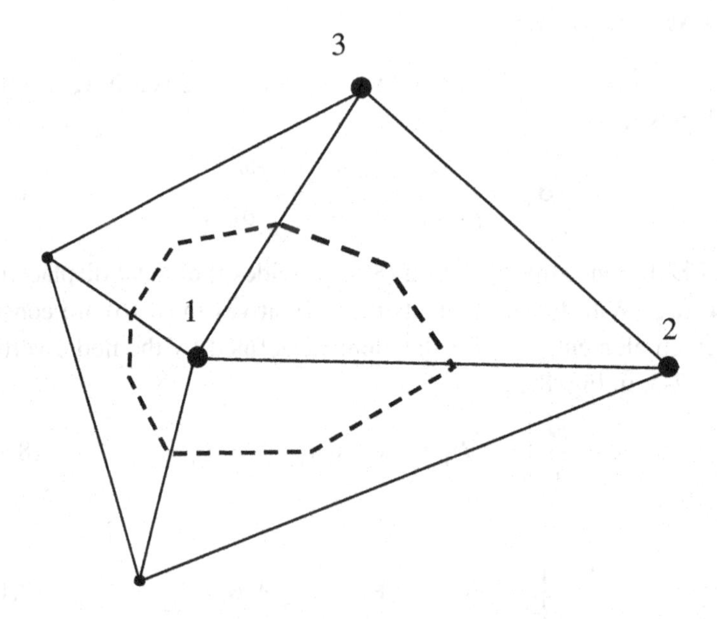

Fig. 8.6 A control volume and element arrangement

point of a face on the control volume (calculated in terms of the element nodal values via the linear shape functions defined in (3.9)), and $\Delta\vec{y}$ is the change in y moving along a given face in a counter-clockwise direction. Equation (8.18) immediately implies that nodal derivatives can be calculated as

$$\left.\frac{\partial\phi}{\partial x}\right|_1 = \sum_{faces}\phi_{mid}\Delta\vec{y}/V_1, \quad \left.\frac{\partial\phi}{\partial y}\right|_1 = -\sum_{faces}\phi_{mid}\Delta\vec{x}/V_1, \tag{8.19}$$

Note as written (8.19) is valid for internal nodes. For boundary nodes the control volume surface includes both faces and the parts of the element sides that coincide with the boundary. In this case (8.19) may require additional terms to account for contributions from the element side surfaces. For example, consider the section of domain boundary (heavy lines), shown in Fig. 8.7. When calculating a derivative associated with the corner node 1, in addition to the two mid face (*a* and *b*) terms in (8.19), the additional terms

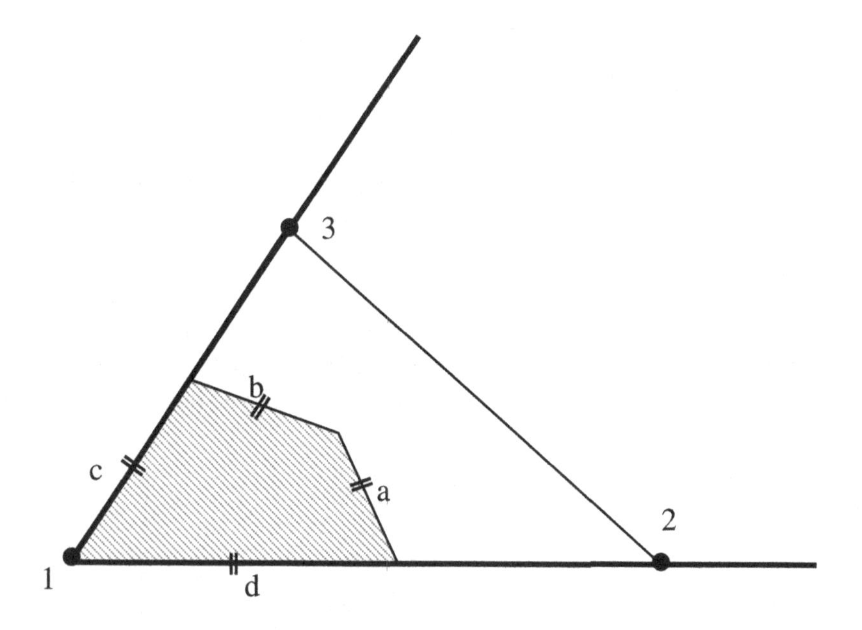

Fig. 8.7 Single element support on a domain boundary

$$
\begin{aligned}
E1 &= \left(\frac{3\phi_1}{4} + \frac{\phi_2}{4}\right)\left(\frac{y_2}{2} - \frac{y_1}{2}\right) \\
E2 &= \left(\frac{3\phi_1}{4} + \frac{\phi_3}{4}\right)\left(\frac{y_1}{2} - \frac{y_3}{2}\right)
\end{aligned}
\qquad (8.20)
$$

evaluated at the mid-points of the element sides (*c* and *d*) need to be included in the calculation of $(\partial\phi/\partial x)_1$. Equation (8.20) can also be written in terms the derivatives of the shape functions, i.e.,

$$
\begin{aligned}
E1 &= -\left(\frac{3\phi_1}{8} N_{3x} + \frac{\phi_2}{8} N_{3x}\right) \\
E2 &= -\left(\frac{3\phi_1}{4} N_{2x} + \frac{\phi_3}{4} N_{2x}\right)
\end{aligned}
\qquad (8.21)
$$

Note, for the elements in the support of an internal node, as we move counter-clockwise from element e1 to element e2

$$N_{2x}^{e1} = -N_{3x}^{e2}, \quad \phi_3^{e1} = -\phi_2^{e2} \tag{8.22}$$

Hence on completing the circuit around node 1 the extra terms in (8.20) or (8.21) will cancel out and the derivatives at the internal node 1 can be calculated directly by (8.19).

An alternative form of (8.19) can be obtained by expanding. Consider the element with vertices labeled 1, 2 and 3 in the support of node 1 in Fig. 8.6. By (8.19) the contributions of the two control volume faces $(f1, f2)$ in this element to the nodal derivative $(\partial\phi/\partial x)_1$ can be expanded as

$$\begin{aligned}\phi_{f1}\Delta\vec{y}_{f1} + \phi_{f2}\Delta\vec{y}_{f2} &= \left(\frac{5\phi_1}{12} + \frac{5\phi_2}{12} + \frac{2\phi_3}{12}\right)\left(-\frac{y_1}{6} - \frac{y_2}{6} + \frac{y_3}{3}\right) \\ &+ \left(\frac{5\phi_1}{12} + \frac{2\phi_2}{12} + \frac{5\phi_3}{12}\right)\left(\frac{y_1}{6} - \frac{y_2}{3} + \frac{y_3}{6}\right)\end{aligned} \tag{8.23}$$

On multiplying out the right hand side, gathering terms and using the definitions of the shape function derivatives

$$\begin{aligned}\phi_{f1}\Delta\vec{y}_{f1} + \phi_{f2}\Delta\vec{y}_{f2} &= \frac{V^{ele}}{3}(N_{1x}\phi_1 + N_{2x}\phi_2 + N_{3x}\phi_3) - E1 - E2 \\ &= \frac{V^{ele}}{3}\left.\frac{\partial\phi}{\partial x}\right|_{ele} - E1 - E2\end{aligned} \tag{8.24}$$

On substitution into (8.19), at internal nodes the extra terms $E1, E2$ in (8.24) cancel by (8.22) and at boundary nodes they cancel by (8.21). Hence, the derivatives at a node can be determined in terms of the area averaged element derivatives, i.e.,

$$\left.\frac{\partial\phi}{\partial x}\right|_1 = \frac{\sum\limits_{\text{support}} \frac{V^{ele}}{3}\left.\frac{\partial\phi}{\partial x}\right|_{ele}}{V_1}, \quad \left.\frac{\partial\phi}{\partial y}\right|_1 = \frac{\sum\limits_{\text{support}} \frac{V^{ele}}{3}\left.\frac{\partial\phi}{\partial y}\right|_{ele}}{V_1} \tag{8.25}$$

an equivalent easier to use form of (8.19).

8.4.3 Estimation of the nodal stress field

From the result in (8.25) the nodal stress field is readily calculated by

$$\sigma_{xx}\Big|_i = \frac{E}{(1-\nu^2)}\left[\frac{\displaystyle\sum_{i^{\text{th}}\text{support}} \frac{V^{ele}}{3}\frac{\partial u_x}{\partial x}\bigg|_{ele}}{V_i} + \nu\frac{\displaystyle\sum_{i^{\text{th}}\text{support}} \frac{V^{ele}}{3}\frac{\partial u_y}{\partial y}\bigg|_{ele}}{V_i}\right] \tag{8.26}$$

With this formula the nodal values of the stress concentration σ_{xx}/σ_0 stress along the boundary $x=0$ can be calculated and compared with the analytical solution in (8.1). This result is shown in Fig. 8.8.

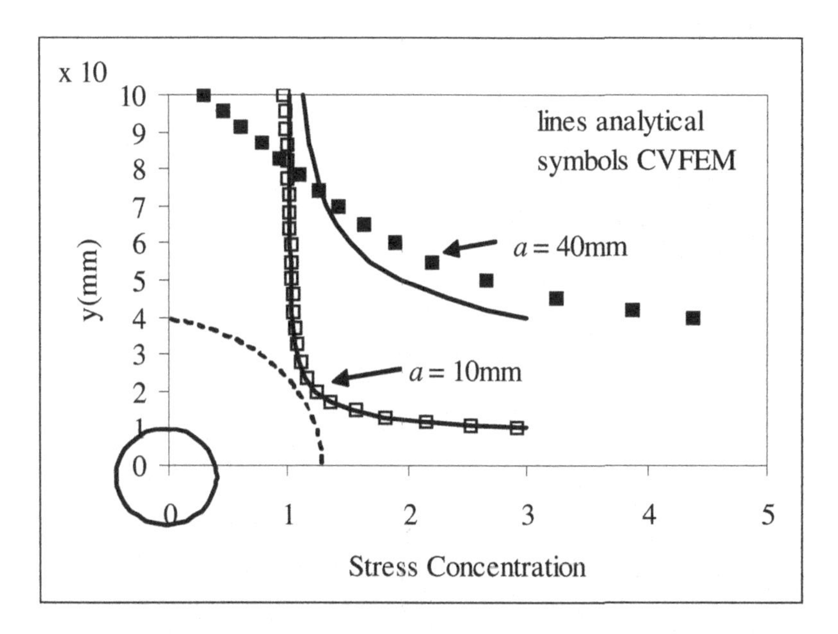

Fig. 8.8 predicted stress on $x = 0$ (symbols) compared with analytical values (8.1)

Note that with a small hole $a=10mm$ the finite plate (200mm×200mm) used in the numerical solution is a reasonable approximation of the infinite plate condition needed for solution (8.1); the analytical and predicted stress concentrations on $x=0$ are very close. When a bigger hole is used however ($a = 40mm$) the validity of the numerical solution to match the analytical conditions clearly breaks down.

8.5 Summary

Although a relatively simple plane stress problem has been solved this chapter contains all of the ingredients required for a reader to develop a CVFEM solution for any plane-stress or plane strain linear elastic problem.

Chapter 9

CVFEM Stream Function-Vorticity Solution for a Lid Driven Cavity Flow

A Control Volume Finite Element Solution for the stream-function—vorticity formulation of lid driven cavity flow is presented.

9.1 Introduction

A classic test problem in computational fluid mechanics is a lid driven cavity flow. In this two dimensional problem, the flow in a square conduit is induced by sliding the upper surface (the lid) at a constant x-velocity U, see Figure 9.1

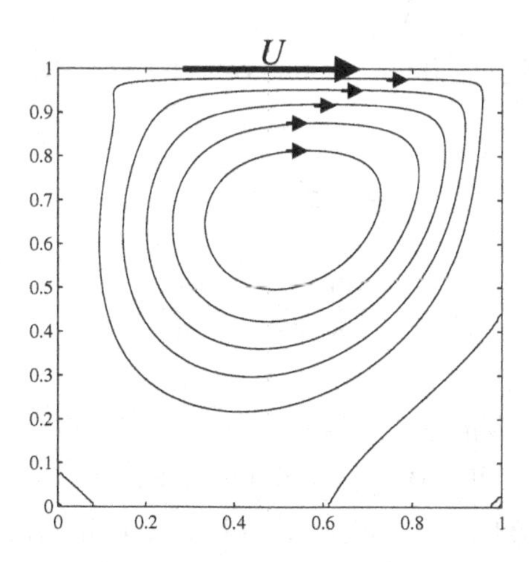

Fig. 9.1 Lid driven cavity flow with streamline indicating flow

The nature of the flow in the cavity is controlled by a Reynolds Number defined as

$$Re = \frac{UL}{\nu} \tag{9.1}$$

Where L is the dimension of the cavity and ν is the kinematic viscosity. Although no analytical solution exists there is an accepted benchmark by Ghia, Ghia and Shin (1982) who use a multi-grid solution of the Navier-Stokes equations in terms of primitive variables (velocity and pressure). In this work a CVFEM stream-function-vorticity solution will be sough.

9.2 The Governing Equations

In terms of the stream function Ψ and vorticity ω the governing equations, in integral form, are given by (2.50) and (2.51),

$$\oint_S \omega \boldsymbol{v} \cdot \boldsymbol{n} - \nu \nabla \omega \cdot \boldsymbol{n} dS = 0 \tag{9.2}$$

$$-\int_A \omega \, dA = \oint_S \nabla \psi \cdot \boldsymbol{n} \, dS \tag{9.3}$$

In terms of the velocity field, $\boldsymbol{v}$, the vorticity and stream function are defined by

$$\omega = \frac{\partial v_y}{\partial x} - \frac{\partial v_x}{\partial y} \tag{9.4}$$

$$v_x = \frac{\partial \Psi}{\partial y}, \; v_y = -\frac{\partial \Psi}{\partial x} \tag{9.5}$$

The boundary condition on (9.3) is

$$\Psi = 0 \text{ on all solid boundaries} \tag{9.6}$$

There is not an explicit boundary condition for the vorticity equation (9.2). As will be detailed below, in setting up the numerical solution of (9.2) and (9.3) a boundary condition for the solution of the discrete form of (9.2) is established by using the discrete form of (9.3) at boundary nodes coupled with the known velocity conditions, i.e.,

$$\begin{aligned} &u_x = U \text{ on lid, } u_x = 0 \text{ on all other boundaries} \\ &u_y = 0 \text{ on all boundaries} \end{aligned} \tag{9.7}$$

9.3 The CVFEM Discretization of the Stream Function Equation

In keeping with rest of this monograph the development of the discreet equations is based on linear triangular elements. The key components of this discretization, the support, control volume, the element, and the control volume faces (*f1*, *f2*) in an element are illustrated in Fig. 9.2. It is convenient to first consider the discretization of the stream function equation (9.3). The form of this equation can be identified as steady state diffusion with a volume source. As such, developing the CVFE discretization can be directly based on the calculations presented in §5.3 and § 5.5. The general form for the discrete equation, for node i in the support shown in Fig. 9.2, is

$$[\, a_i^{\Psi} + Q_{C_i}^{\Psi} + B_{C_i}^{\Psi} \,]\Psi_i = \sum_{j=1}^{n_i} a_{i,j}^{\Psi} \Psi_{S_{i,j}} + Q_{B_i}^{\Psi} + B_{B_i}^{\Psi} \tag{9.8}$$

where the a's are the coefficients, the index (i,j) indicates the jth node ($j = 1,2,..n_i$) in the support of node i, the index $S_{i,j}$ provides the global node number ($i = 1,2,3,..n$) of the jth node in the support, the B's account for boundary conditions, and the Q 's for source terms.

9.3.1 Diffusion contributions

Following the presentation in §5.3 the contribution to the coefficients a in (9.8), obtained from consideration of the diffusion flux across the control volume faces (*f1*, *f2*) for the particular element in Fig. 9.2, are

$$
\begin{aligned}
a_1^{\kappa} &= -N_{1x}\Delta\vec{y}_{f1} + N_{1y}\Delta\vec{x}_{f1} - N_{1x}\Delta\vec{y}_{f2} + N_{1y}\Delta\vec{x}_{f2} \\
a_2^{\kappa} &= N_{2x}\Delta\vec{y}_{f1} - N_{2y}\Delta\vec{x}_{f1} + N_{2x}\Delta\vec{y}_{f2} - N_{2y}\Delta\vec{x}_{f2} \\
a_3^{\kappa} &= N_{3x}\Delta\vec{y}_{f1} - N_{3y}\Delta\vec{x}_{f1} + N_{3x}\Delta\vec{y}_{f2} - N_{3y}\Delta\vec{x}_{f2}
\end{aligned}
\tag{9.9}
$$

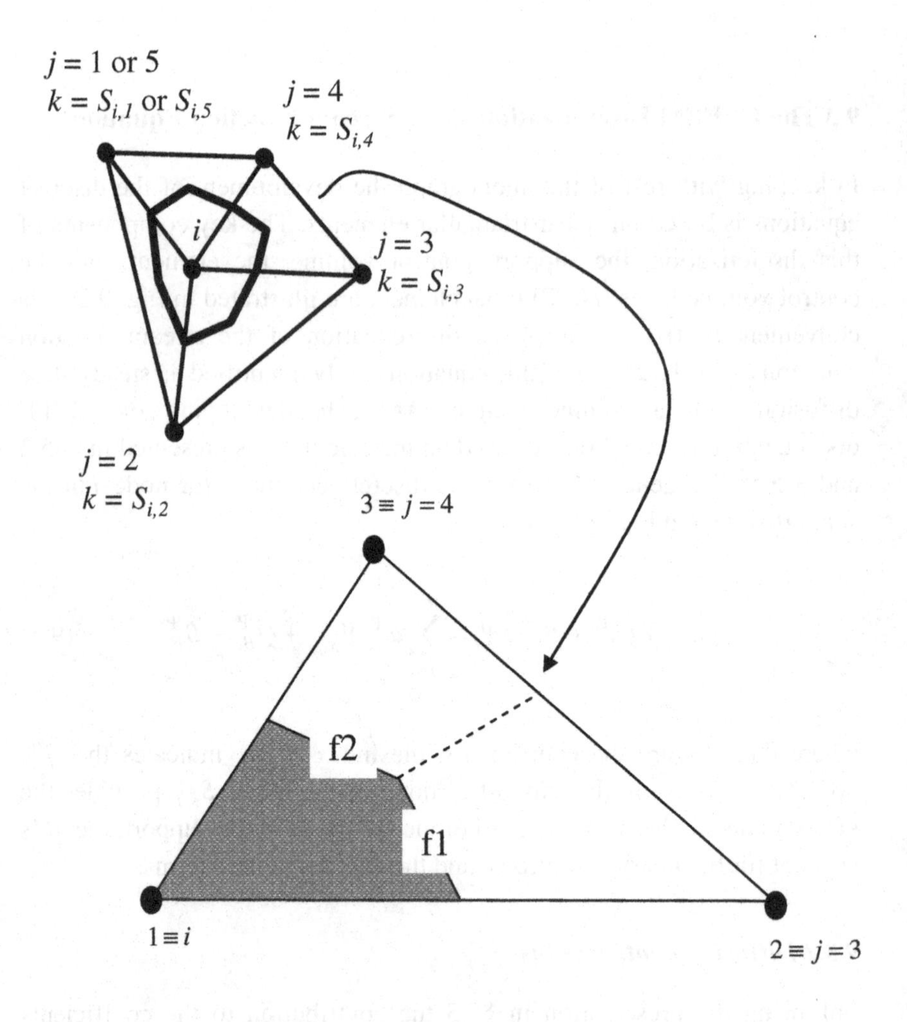

Fig. 9.2 Arrangement of control volumes and elements

These values can be used to *update* the ith support coefficients through

$$\begin{aligned} a_i^{\Psi} &= a_i^{\Psi} + a_1^{\kappa} \\ a_{i,3}^{\Psi} &= a_{i,3}^{\Psi} + a_2^{\kappa} \\ a_{i,4}^{\Psi} &= a_{i,4}^{\Psi} + a_4^{\kappa} \end{aligned} \tag{9.10}$$

In (9.9), moving counter-clockwise around node i, the signed distances

$$\Delta\vec{x}_{f1} = \frac{x_3}{3} - \frac{x_2}{6} - \frac{x_1}{6}, \Delta\vec{x}_{f2} = -\frac{x_2}{3} + \frac{x_3}{6} + \frac{x_1}{6} \tag{9.11a}$$

$$\Delta\vec{y}_{f1} = \frac{y_3}{3} - \frac{y_2}{6} - \frac{y_1}{6}, \quad \Delta\vec{y}_{f2} = -\frac{y_2}{3} + \frac{y_3}{6} + \frac{y_1}{6} \tag{9.11b}$$

the derivatives of the shape functions

$$\begin{aligned} N_{1x} &= \frac{\partial N_1}{\partial x} = \frac{(y_2 - y_3)}{2V^{ele}}, \quad N_{1y} = \frac{\partial N_1}{\partial y} = \frac{(x_3 - x_2)}{2V^{ele}} \\ N_{2x} &= \frac{\partial N_2}{\partial x} = \frac{(y_3 - y_1)}{2V^{ele}}, \quad N_{2y} = \frac{\partial N_2}{\partial y} = \frac{(x_1 - x_3)}{2V^{ele}} \\ N_{3x} &= \frac{\partial N_3}{\partial x} = \frac{(y_1 - y_2)}{2V^{ele}}, \quad N_{3y} = \frac{\partial N_3}{\partial y} = \frac{(x_2 - x_1)}{2V^{ele}} \end{aligned} \tag{9.12}$$

and the volume of the element

$$V^{ele} = \frac{(x_2 y_3 - x_3 y_2) + x_1(y_2 - y_3) + y_1(x_3 - x_2)}{2} \tag{9.13}$$

9.3.2 Source terms

The left hand side of (9.3) is identified as a volume source term which can be represented in (9.8) by the setting

$$Q_{B_i}^{\Psi} = \omega_i V_i, \qquad Q_{C_i}^{\Psi} = 0 \tag{9.14}$$

where V_i is the volume of the control volume, made up of contributions of $\frac{1}{3}V^{ele}$ from each element in the support of node i. The coupling with the nodal vorticity field will require an iterative solution of (9.8). At each iteration, the source term in (9.14) will need to be re-evaluated to reflect the updating of the iterative values of the nodal vorticity field.

9.3.3 Boundary conditions

The boundary conditions of $\Psi = 0$ on all boundaries are set in (9.8) by setting, for each node i lying on the domain boundary

$$B_{B_i}^{\Psi} = 0, \qquad B_{C_i}^{\Psi} = 10^{15} \tag{9.15}$$

9.4 The CVFEM Discretization of the Vorticity Equation

The vorticity equation (9.2) is seen to have the form of a steady state advection diffusion equation for a scalar. As such, the discreet form can be obtained directly from calculations presented in §5.3 and § 5.4 of Chapter 5. The general form for this equation for node i in the support shown in Fig. 9.2 is

$$[\, a_i^{\omega} + B_{C_i}^{\omega} \,]\omega_i = \sum_{j=1}^{n_i} a_{i,j}^{\omega} \omega_{S_{i,j}} + B_{B_i}^{\omega} \tag{9.16}$$

9.4.1 Diffusion contributions

Following the presentation in §5.3 the contribution to the coefficients a in (9.16), from consideration of the diffusion flux across the control volume faces (*f1*, *f2*) in Fig. 9.2, are

$$\begin{aligned}
a_1^{\kappa} &= -\nu N_{1x} \Delta\vec{y}_{f1} + \nu N_{1y} \Delta\vec{x}_{f1} - \nu N_{1x} \Delta\vec{y}_{f2} + \nu N_{1y} \Delta\vec{x}_{f2} \\
a_2^{\kappa} &= \nu N_{2x} \Delta\vec{y}_{f1} - \nu N_{2y} \Delta\vec{x}_{f1} + \nu N_{2x} \Delta\vec{y}_{f2} - \nu N_{2y} \Delta\vec{x}_{f2} \\
a_3^{\kappa} &= \nu N_{3x} \Delta\vec{y}_{f1} - \nu N_{3y} \Delta\vec{x}_{f1} + \nu N_{3x} \Delta\vec{y}_{f2} - \nu N_{3y} \Delta\vec{x}_{f2}
\end{aligned} \tag{9.17}$$

These values can be used to *update* the ith support coefficients through

$$a_i^{\omega} = a_i^{\omega} + a_1^{\kappa}$$
$$a_{i,3}^{\omega} = a_{i,3}^{\omega} + a_2^{\kappa} \qquad (9.18)$$
$$a_{i,4}^{\omega} = a_{i,4}^{\omega} + a_4^{\kappa}$$

9.4.2 The advection coefficients

Following the presentation in §5.4 the contribution to the coefficients a in (9.8), from consideration of the advection flux across the control volume faces (*f1, f2*) in Fig. 9.2, are, using upwinding,

$$a_1^u = \max[q_{f1},0] + \max[q_{f2},0]$$
$$a_2^u = \max[-q_{f1},0] \qquad (9.19)$$
$$a_3^u = \max[-q_{f2},0]$$

These values can be used to *update* the ith support coefficients through

$$a_i^{\omega} = a_i^{\omega} + a_1^{\kappa} + a_1^u$$
$$a_{i,3}^{\omega} = a_{i,j}^{\omega} + a_2^{\kappa} + a_2^u \qquad (9.20)$$
$$a_{i,4}^{\omega} = a_{i,j}^{\omega} + a_3^{\kappa} + a_3^u$$

In (9.19) the volume flow rates out across the control volume faces are calculated by

$$q_{f1} = \boldsymbol{v} \cdot \boldsymbol{n}A\big|_{f1} = v_x^e \Delta\vec{y}_{f1} - v_y^e \Delta\vec{x}_{f1} \qquad (9.21a)$$

$$q_{f2} = \boldsymbol{v} \cdot \boldsymbol{n}A\big|_{f2} = v_x^e \Delta\vec{y}_{f2} - v_y^e \Delta\vec{x}_{f2} \qquad (9.21b)$$

where the velocities at the mid point of the faces are approximated in terms of the constant element velocity, $\boldsymbol{v}^e = (v_x^e, v_y^e)$. This value, in turn is calculated directly from the nodal stream function field through the approximation of (9.5), e.g., for the element in Fig. 9.2

$$
\begin{aligned}
v_x^e &\approx N_{1y}\Psi_i + N_{2y}\Psi_{S_{i,3}} + N_{3y}\Psi_{S_{i,4}} \\
v_y^e &\approx -N_{1x}\Psi_i - N_{2x}\Psi_{S_{i,3}} - N_{3x}\Psi_{S_{i,4}}
\end{aligned}
\tag{9.22}
$$

This coupling with the nodal stream-function field will require an iterative solution of (9.16). At each iteration, the advection coefficients in (9.19) will need to be evaluated to reflect the updating of the iterative values of the nodal stream-function field.

9.4.3 Boundary conditions

Before (9.16) can be solved boundary conditions need to be prescribed. Fixed value boundary conditions are used. At each node i on a domain boundary the discrete form of the stream function equation (9.3) can be used to prescribe a value for the nodal vorticity. Use of this equation is allowed since, due to the fact that the stream function takes the known fixed value $\Psi = 0$ on all boundaries, it is not needed in the stream function solution. At a node i on the boundary the control volume finite element discretization of (9.3), informed by (9.8), can be written as

$$
\omega_i = \frac{1}{V_i}\left[a_i^{\Psi}\Psi_i - \sum_{j=1}^{n_i} a_{i,j}^{\Psi}\Psi_{S_{i,j}} - \sum_{bounday} A\nabla\Psi \cdot \boldsymbol{n} \right]
\tag{9.23}
$$

where the coefficients a^{Ψ} are given by (9.10) and the definition of the source term in (9.14) has been used to isolate ω_i. The last term in (9.23) represents contributions from control volume faces that coincide with a boundary segment, see the double line in Fig 9.3. In the sliding lid problem of Fig. 9.1 the only non-zero contribution from this term will be for nodes on the sliding lid and will have the form

$$
\left[\sum_{bounday} A\nabla\Psi \cdot \boldsymbol{n} \right] = \Delta U
\tag{9.24}
$$

where Δ is the length of the control volume surface on the boundary segment, see Fig. 9.3.

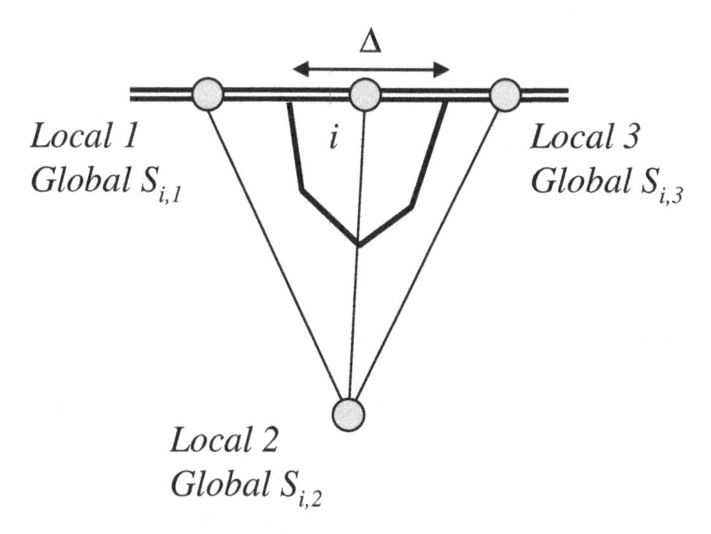

Fig. 9.3 Node point on sliding lid boundary

Hence, with current iterative values of the stream function known, (9.23) and (9.24) can be used to estimate fixed nodal values of vorticity for boundary nodes i

$$\omega_i = \frac{1}{V_i}\left[a_i^{\Psi}\Psi_i - \sum_{j=1}^{n_i} a_{i,j}^{\Psi}\Psi_{S_{i,j}} - \begin{cases} \Delta_i U & i \in lid \\ 0 & \text{otherwise} \end{cases} \right] \tag{9.25}$$

These values are forced into the solution of (9.16) by setting

$$B_{B_i}^{\omega} = \omega_i \times 10^{15}, \qquad B_{C_i}^{\omega} = 10^{15}$$

9.5 Solution Steps

9.5.1 Nested iteration

As noted an iterative solution of the coupled equations (9.8) and (9.16) is required. This is achieved with a nested iteration loop.

An outer iteration is made *1- outer* (~1000)

With the current best estimates the boundary values of ω_i are obtained from (9.25)

The coefficients in (9.16) are calculated.

Equation (9.16) is solved by a simple point iteration scheme; 20 inner iteration steps are used.

The coefficients for (9.8) are calculated; the latest iterative values of the vorticity are used.

Equation (9.8) is solved by a simple point iteration scheme; 20 inner iteration steps are used.

The default choice of 1000 iterations is usually enough to ensure convergence. A more sophisticated approach could use a convergence check to determine the outer iteration closure.

9.5.2 Calculating the nodal velocity field

Following the calculation of the converged Ψ field the nodal velocity field $\boldsymbol{v}$ can be calculated. Integrating (9.5) over the ith control volume results in

$$\int_{V_i} v_x dV = \int_{V_i} \frac{\partial \Psi}{\partial y} dV, \qquad \int_{V_i} v_y dV = -\int_{V_i} \frac{\partial \Psi}{\partial x} dV, \tag{9.26}$$

or using the divergence theorem

$$\int_{V_i} v_x dV = \oint_S \Psi n_y dS, \qquad \int_{V_i} v_y dV = -\oint_S \Psi n_x \tag{9.27}$$

where the surface integral is over the control volume faces, and $\boldsymbol{n} = (n_x, n_y)$ is the outward pointing normal on the control volume face. Using mid point integration approximations

$$v_{x_i} = \frac{1}{V_i} \sum_{faces} \Psi A n_y, \qquad v_{y_i} = -\frac{1}{V_i} \sum_{faces} \Psi A n_x \tag{9.28}$$

where A is the area (length × unit depth) of a given control volume face, and , since boundary values of $\Psi = 0$, the summations are restricted to control faces that do not coincide with the domain boundaries. On recalling that the outward normal area product on a volume face is

$$An_x = \Delta\vec{y},\ An_y = -\Delta\vec{x}$$

see (9.11) for definitions of $\Delta\vec{y}$ and $\Delta\vec{x}$, the contributions to control volume i from the right hand side of (9.28), for the selected element from the control volume in Fig. 9.2, can be calculated as

$$\begin{aligned}
\sum_{f1,f2} \Psi A n_y &= -\left(\tfrac{5}{12}\Psi_i + \tfrac{5}{12}\Psi_{S_{i,3}} + \tfrac{2}{12}\Psi_{S_{i,4}}\right)\Delta\vec{x}_{f1} - \\
&\qquad \left(\tfrac{5}{12}\Psi_i + \tfrac{2}{12}\Psi_{S_{i,3}} + \tfrac{5}{12}\Psi_{S_{i,4}}\right)\Delta\vec{x}_{f2} \\
\sum_{f1,f2} \Psi A n_x &= -\left(\tfrac{5}{12}\Psi_i + \tfrac{5}{12}\Psi_{S_{i,3}} + \tfrac{2}{12}\Psi_{S_{i,4}}\right)\Delta\vec{y}_{f1} - \\
&\qquad \left(\tfrac{5}{12}\Psi_i + \tfrac{2}{12}\Psi_{S_{i,3}} + \tfrac{5}{12}\Psi_{S_{i,4}}\right)\Delta\vec{y}_{f2}
\end{aligned} \tag{9.29}$$

With the nodal velocity field calculated from (9.28) the velocity at given point $\boldsymbol{x} = (x, y)$ can be determine by determining the triangular element (local nodes 1,2, and 3) which contains the point $\boldsymbol{x}$ and then using the shape function interpolations

$$\begin{aligned}
v_x(\boldsymbol{x}) &= N_1 v_{x_1} + N_2 v_{x_2} + N_3 v_{x_3} \\
v_y(\boldsymbol{x}) &= N_1 v_{y_1} + N_2 v_{y_2} + N_3 v_{y_3}
\end{aligned} \tag{9.30}$$

Where the shape function N_i $(i = 1,2,3)$ are defined in (3.7) and (3.8).

9.6 Results

For all results, unless otherwise stated, the settings $U = 1, L = 1$ will be made; in this way the Reynolds number of the Flow will be $Re = 1/\nu$. The calculated flow-field, given by the stream-lines, when $Re = 400$ is shown in Fig. 9.1. This calculation was carried out on a structured mesh with 61 rows and 61 columns; a section of this grid is shown in Fig 9.4.

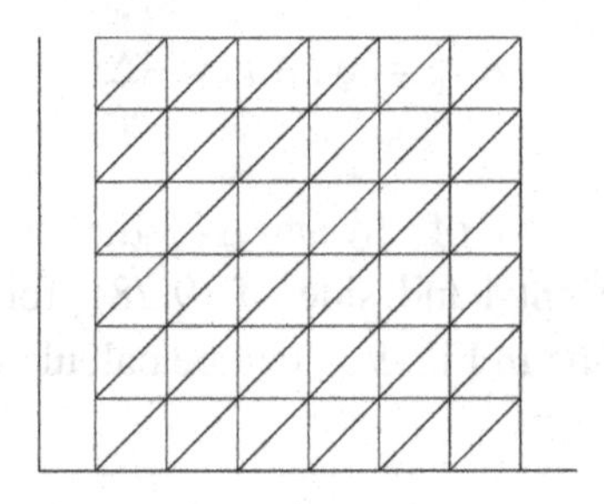

Fig. 9.4 Section of structured 61×61 grid

Ghia, Ghia and Shin (1982) tabulate the benchmark v_x velocity along the line x = 0.5 and the v_y velocity along the line y = 0.5 for Reynolds numbers of $Re = 100$ and 400. Fig. 9.5 compares the tabulated $Re = 100$ velocities with the calculated velocities based on the CVFEM stream-function—vorticity solution on the structured grid of Fig. 9.4. The agreement between the CVFEM and the benchmark is close. Fig. 9.6 compares the tabulated $Re = 400$ velocities with the calculated velocities based on the CVFEM stream-function—vorticity solution on the structured grid of Fig. 9.4. In this case, the agreement between the CVFEM and the benchmark, although still reasonable, is not as close as the $Re = 100$ case. This indicates that a finer grid is needed or a more sophisticated treatment of the vorticity boundary condition.

In contrast to the structured grid results of Figs. 9.1, 9.5 and 9.6, the CVFEM stream-function—vorticity solution on the unstructured mesh of Fig. 9.7, for $Re = 100$, are compared with the tabulated benchmark in Fig. 9.8. The quality of these predictions is consistent with the structured grid values of Fig. 9.5.

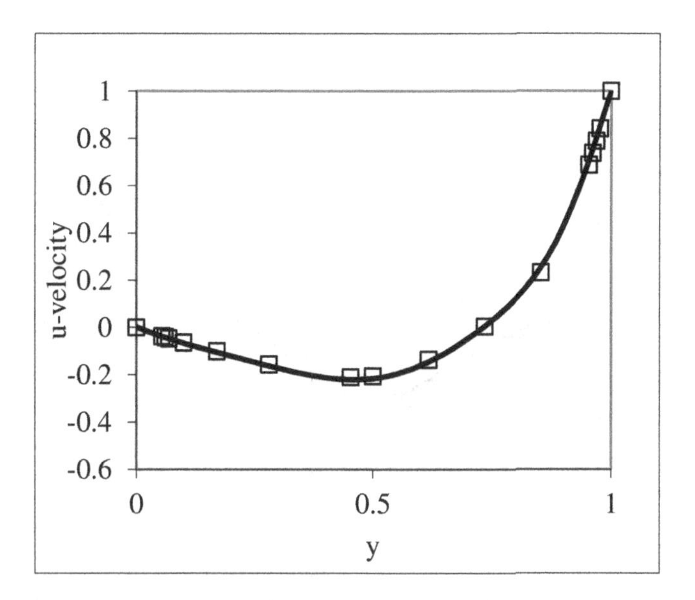

Fig. 9.5a u-velocity in sliding lid cavity along x = 0.5. Re =100, structured grid of Fig. 9.4. Symbols benchmark solution Ghia et al. (1982), lines CVFEM solution

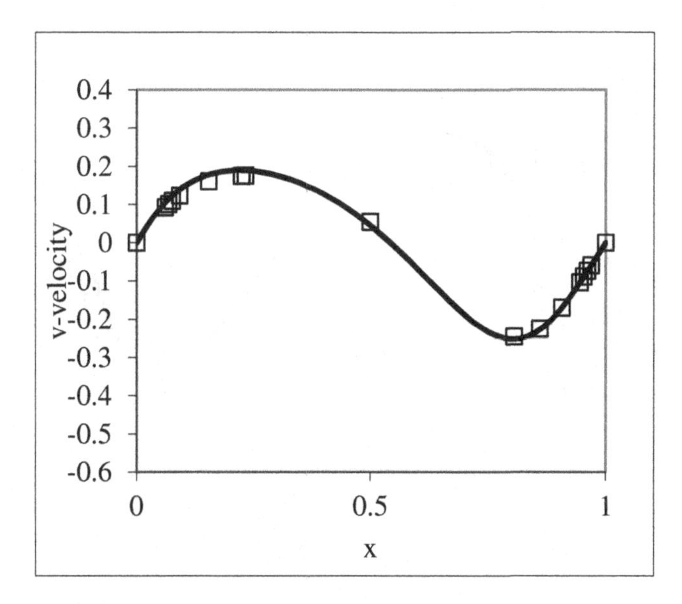

Fig. 9.5b v-velocity in sliding lid cavity along y = 0.5. Re =100, structured grid of Fig. 9.4. Symbols benchmark solution Ghia et al. (1982), lines CVFEM solution

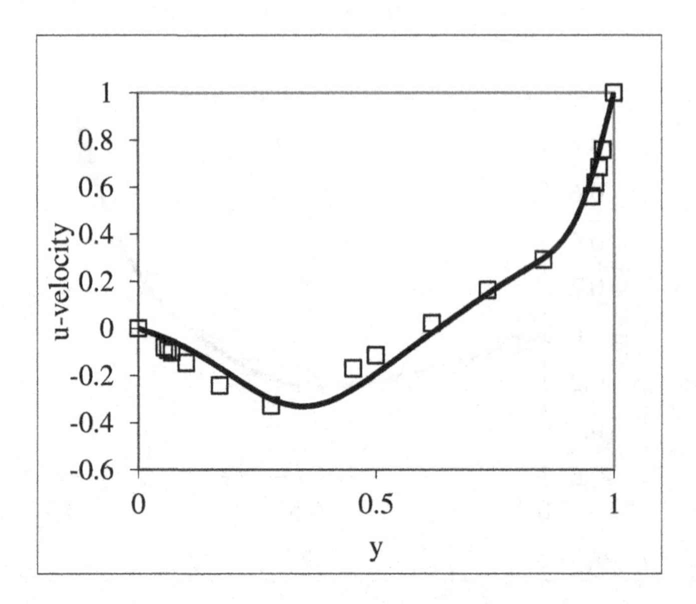

Fig. 9.6a u-velocity in sliding lid cavity along x = 0.5. Re =400, structured grid of Fig. 9.4. Symbols benchmark solution Ghia et al. (1982), lines CVFEM solution.

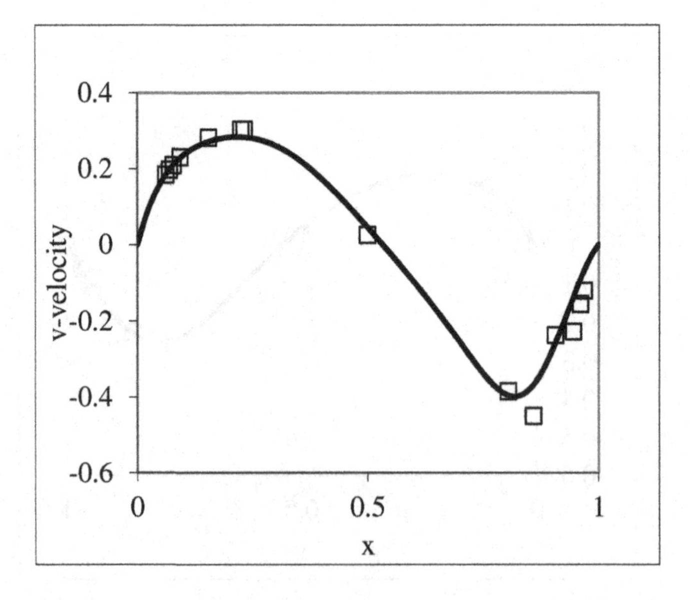

Fig. 9.6b v-velocity in sliding lid cavity along y = 0.5 (Re = 400, structured grid) Symbols benchmark solution Ghia et al. (1982), lines CVFEM solution.

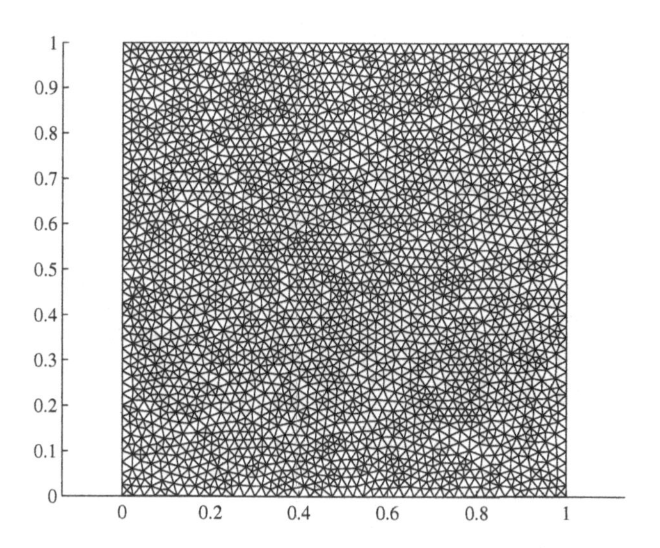

Fig. 9.7 Unstrctured mesh for stream-function—vorticity CVFEM solution

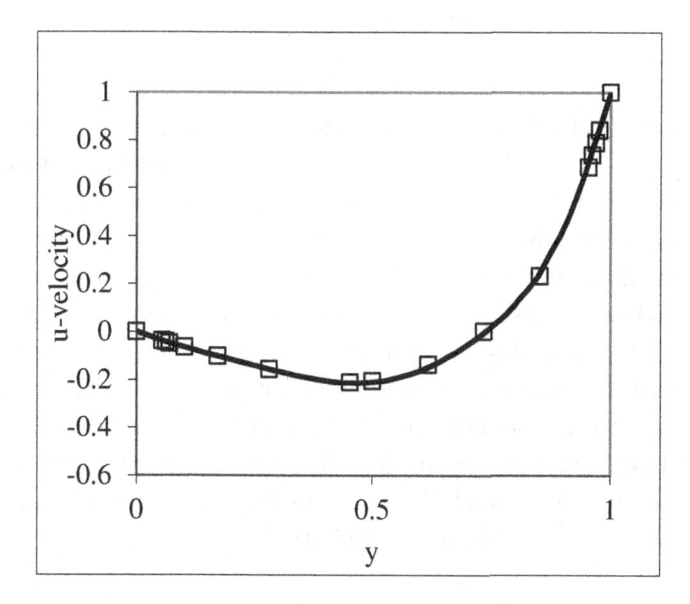

Fig. 9.8a u-velocity in sliding lid cavity along x = 0.5. Re =100, unstructured grid of Fig. 9.7 Symbols benchmark solution Ghia et al. (1982), lines CVFEM solution

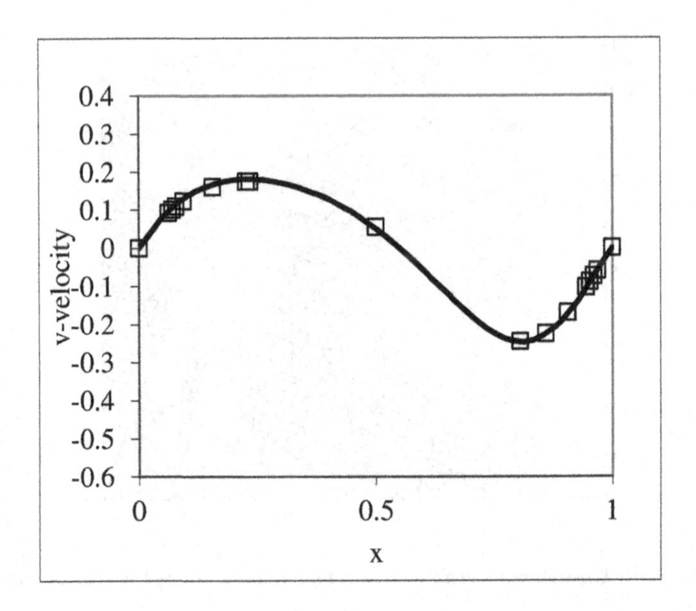

Fig. 9.8b v-velocity in sliding lid cavity along y = 0.5. Re = 100, unstructured grid of Fig. 9.7 Symbols benchmark solution Ghia et al. (1982), lines CVFEM solution

Since the CVFEM solution can fit arbitrary geometries it is interesting to apply it to a modification of the sliding lid problem. One possible variation is to use a ¼ annulus cavity. An unstructured mesh and geometry for this problem is shown in Fig. 9.9. Flow in the cavity is generated by sliding the top lid at a constant x-velocity of $u_x = 1$. Using the thickness of the annulus ($t = 1$) as the length the viscosity is chosen to set the Reynolds number at $Re = 100$. The predictions from the CVFEM stream-function—vorticity solution are presented as streamlines in Fig. 9.10; the stream lines in this plot are in the clockwise direction (negative) and vary from $\psi = -0.1, -0.08, -0.06, -0.04, -0.02, -0.00001$. This solution is a sound illustration of the power of the CVFEM method to solve complex problems in arbitrary domains.

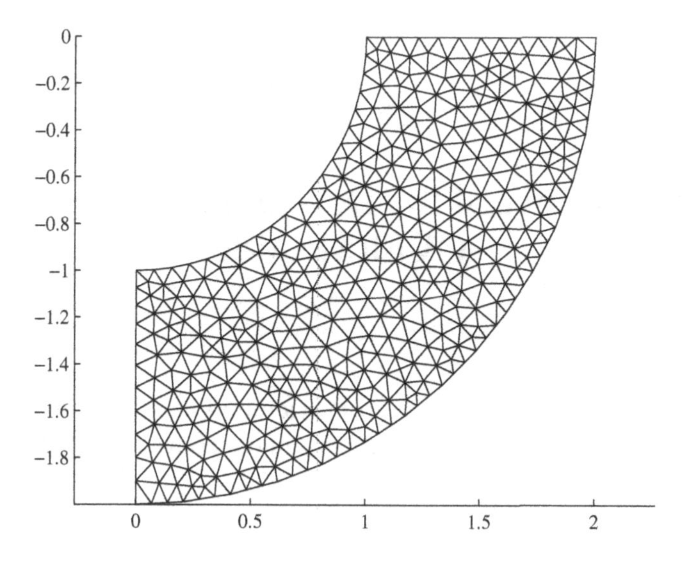

Fig. 9.9 Unstructured grid and geometry for annulus sliding lid problem

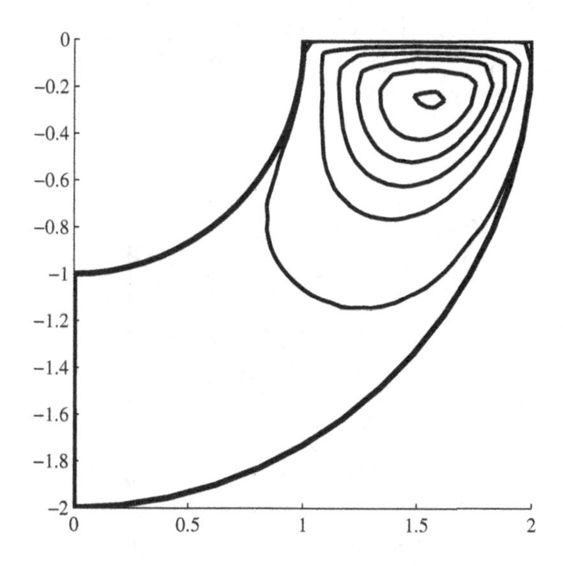

Fig. 9.10 Negative streamlines (~0 to -0.1) predicted by CVFEM for mesh of Fig. 9.9

Chapter 10

Notes toward the Development of a 3-D CVFEM Code

The objective of this chapter is to provide outline notes that will allow the reader to extend the basic concepts laid out above toward the building of a three-dimensional CVFERM code.

10.1 The Tetrahedron Element

If linear approximations are to be used then the element to use in a 3-D CVFEM code is a tetrahedron. A tetrahedron is a three dimensional object with four planar triangular surfaces, Fig 10.1. The vertices of this figure have the Cartesian coordinates (I_x, I_y, I_z), (A_x, A_y, A_z), (B_x, B_y, B_z), and (C_x, C_y, C_z).

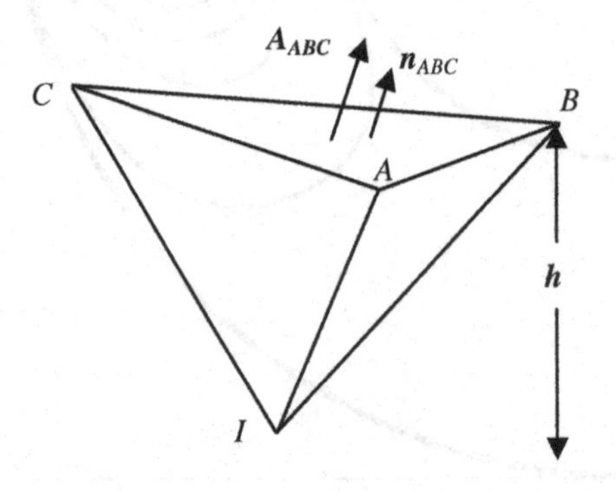

Fig. 10.1 A tetrahedron element

10.2 Creating a Mesh of Tetrahedron Elements

An straightforward way to create a mesh of tetrahedron elements is to start with a three dimensional structured grid of nodes placed at the vertices of plane sided cubes, Fig. 10.2.

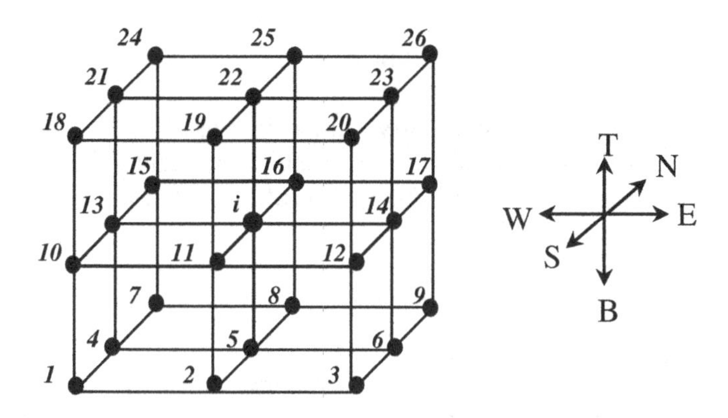

Fig. 10.2 A 3-D grid of nodes placed on the vertices of cube elements

For any give cube element in Fig. 10.2, six (6) tetrahedron elements with nodes at the vertices can be cut. With reference to the cube in Fig. 10.3

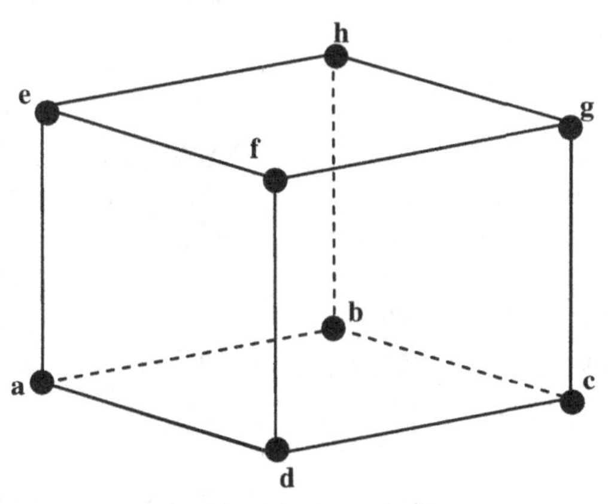

Fig. 10.3 A given cube element with vertex nodes

the six internal, non-overlapping tetrahedrons that can be cut from the cube are made up of the following vertices

a b d e
b f e h
b d e f
c f g h
b c d f
b c f h

These tetrahedrons are explicitly shown in Fig. 10.4.The tetrahedrons shown in this figure can be cut from each of the eight cubes in Fig. 10.2 that are connected to node i. This will result in 24 tetrahedrons that share the common vertex I, viz,

Table 10.1 Tetrahedron in support of node *i* in Fig. 10.2

TNW				BNW			
i	15	13	21	*i*	13	15	7
i	15	22	21	*i*	13	5	7
i	15	16	22	*i*	15	8	16
TNE				*i*	5	7	8
i	22	14	16	*i*	7	15	8
TSW				BNE			
i	19	22	21	*i*	5	6	8
i	19	13	11	*i*	14	8	16
i	19	21	13	*i*	14	6	8
TSE				BSW			
i	19	11	12	*i*	11	13	5
i	19	20	22	BSE			
i	19	20	12	*i*	11	12	5
i	20	12	14	*i*	12	14	6
i	20	14	22	*i*	12	5	6

where, noting the axes labeling in Fig. 10.2, TNE refers to the cube in the Top North East corner. The nodes points listed comprise the support of node *i* and the 24 tetrahedrons are the elements in the support of node *i*.

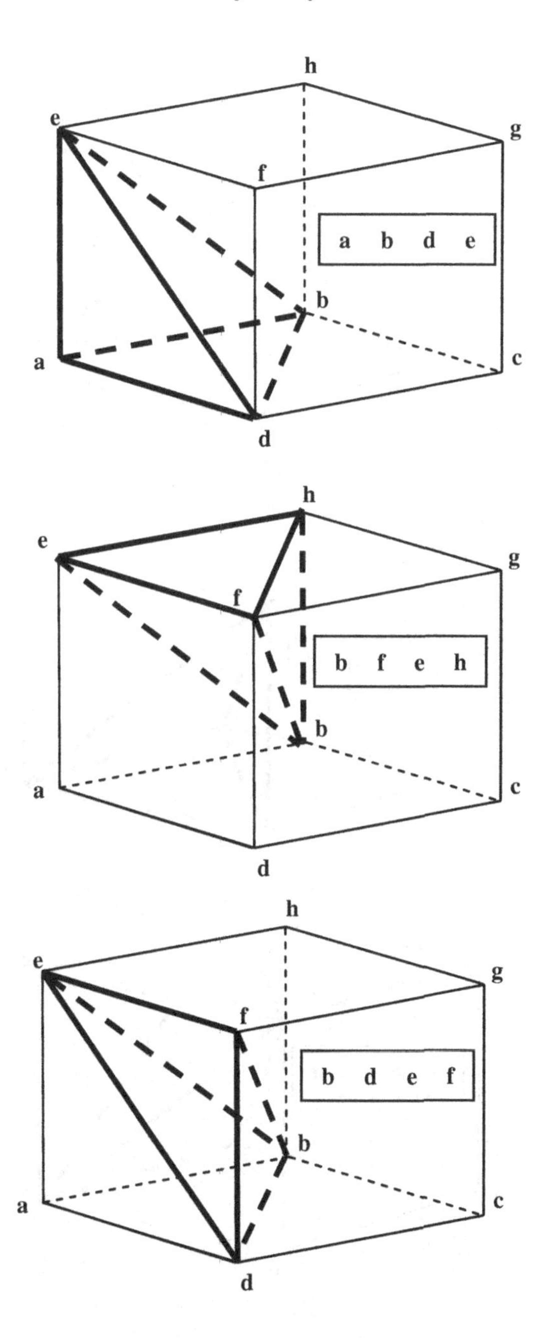

Fig. 10.4a Tetrahedrons cut from left hand side

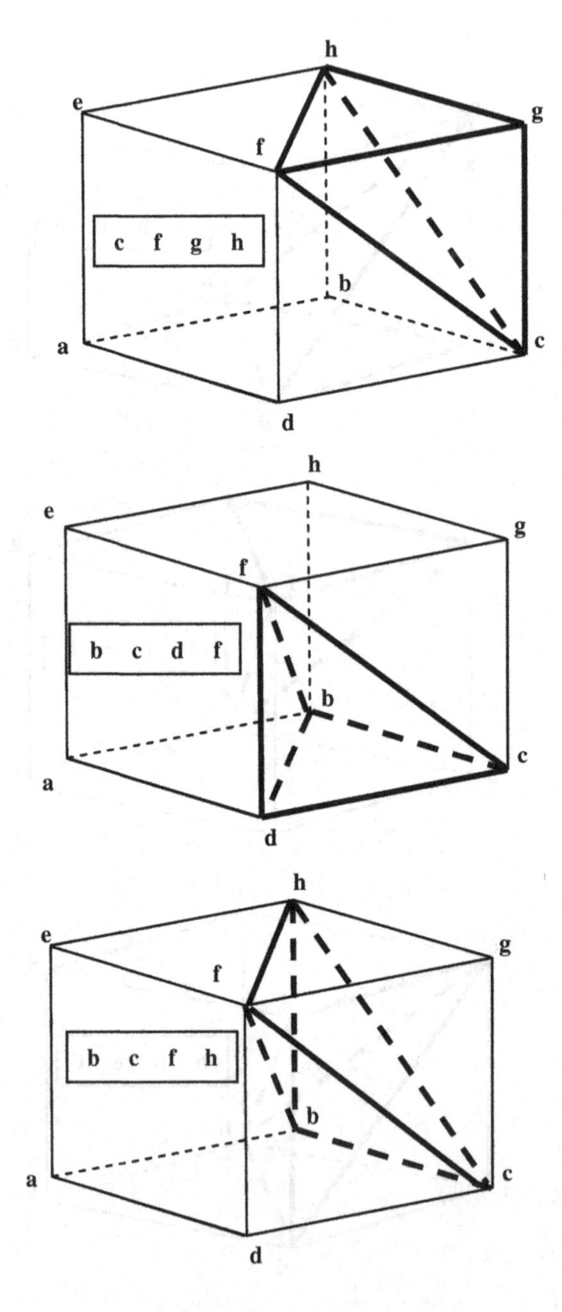

Fig. 10.4b Tetrahedrons cut from right hand side

10.3 Geometric Features of Tetrahedrons

To move forward it is necessary to provide calculations for key geometric features of a tetrahedron. The volume is given by $V = \frac{1}{3}Ah$ where A is the area of a given triangular surface and h is the perpendicular distance from that side to the opposite vertex. In words, with reference to Fig. 10.1,

Volume = $\frac{1}{3}\times$ Area of Surface (*ABC*) $\times$ perpendicular distance from (*ABC*) to *I*

If the vectors from a fixed origin to the vertices of the tetrahedron in Fig 10.1 are

$$\boldsymbol{I} = (I_x, I_y, I_z)$$
$$\boldsymbol{A} = (A_x, A_y, A_z)$$
$$\boldsymbol{B} = (B_x, B_y, C_z)$$
$$\boldsymbol{C} = (C_x, C_y, C_z)$$

(i) the vector product $\frac{1}{2}(\boldsymbol{B}-\boldsymbol{A})\times(\boldsymbol{C}-\boldsymbol{A})$ gives a vector with magnitude equal to the area of the surface ABC and direction normal to the surface ABC, and (ii) the dot product $h = (\boldsymbol{A}-\boldsymbol{I})\cdot \boldsymbol{n}_{ABC} > 0$, where $\boldsymbol{n}_{ABC}$ is the unit normal on ABC that points *out of* tetrahedron $IABC$, gives the height (perpendicular distance) from the surface ABC to vertex *I*. In this way, on reference to the word equation given above the volume of the tetrahedron can be written as

$$\begin{aligned} V = \tfrac{1}{6}|T_I|, \quad \text{where} \quad T_I &= (\boldsymbol{A}-\boldsymbol{I})\cdot((\boldsymbol{B}-\boldsymbol{A})\times(\boldsymbol{C}-\boldsymbol{A})) \\ &= \boldsymbol{A}\cdot(\boldsymbol{B}\times\boldsymbol{C}) \end{aligned} \qquad (10.1a)$$

with a clockwise rotation of the vertices *I*, *A* , *B*, and *C* the volume can also be calculated with any one of the following

$$\begin{aligned} V &= \tfrac{1}{6}|T_A|, \quad \text{where} \quad T_A = (\boldsymbol{B}-\boldsymbol{A})\cdot((\boldsymbol{C}-\boldsymbol{B})\times(\boldsymbol{I}-\boldsymbol{B})) \\ V &= \tfrac{1}{6}|T_B|, \quad \text{where} \quad T_B = (\boldsymbol{C}-\boldsymbol{B})\cdot((\boldsymbol{I}-\boldsymbol{C})\times(\boldsymbol{A}-\boldsymbol{C})) \\ V &= \tfrac{1}{6}|T_C|, \quad \text{where} \quad T_C = (\boldsymbol{I}-\boldsymbol{C})\cdot((\boldsymbol{A}-\boldsymbol{I})\times(\boldsymbol{B}-\boldsymbol{I})) \end{aligned} \qquad (10.1b)$$

For future use, it is also noted that, the vector

$$\boldsymbol{A}_{ABC} = \begin{cases} (\boldsymbol{B}-\boldsymbol{A})\times(\boldsymbol{C}-\boldsymbol{A}), & T_I > 0 \\ -(\boldsymbol{B}-\boldsymbol{A})\times(\boldsymbol{C}-\boldsymbol{A}), & T_I < 0 \end{cases} \tag{10.2a}$$

has magnitude equal to the $2\times$ area of the surface *ABC* and direction along the outward pointing normal $\boldsymbol{n}_{ABC}$. By a clockwise rotation the vectors

$$\boldsymbol{A}_{BCI} = \begin{cases} (\boldsymbol{C}-\boldsymbol{B})\times(\boldsymbol{I}-\boldsymbol{B}), & T_A > 0 \\ -(\boldsymbol{C}-\boldsymbol{B})\times(\boldsymbol{I}-\boldsymbol{B}), & T_A < 0 \end{cases} \tag{10.2b}$$

$$\boldsymbol{A}_{CIA} = \begin{cases} (\boldsymbol{I}-\boldsymbol{C})\times(\boldsymbol{A}-\boldsymbol{C}), & T_B > 0 \\ -(\boldsymbol{I}-\boldsymbol{C})\times(\boldsymbol{A}-\boldsymbol{C}), & T_B < 0 \end{cases} \tag{10.2c}$$

$$\boldsymbol{A}_{IAB} = \begin{cases} (\boldsymbol{A}-\boldsymbol{I})\times(\boldsymbol{B}-\boldsymbol{I}), & T_C > 0 \\ -(\boldsymbol{A}-\boldsymbol{I})\times(\boldsymbol{B}-\boldsymbol{I}), & T_C < 0 \end{cases} \tag{10.2d}$$

are the counterparts for the surfaces *BCI, CIA,* and *IAB* respectively; each vector pointing out of the tetrahedron *IABC* in the direction of the surface unit normals.

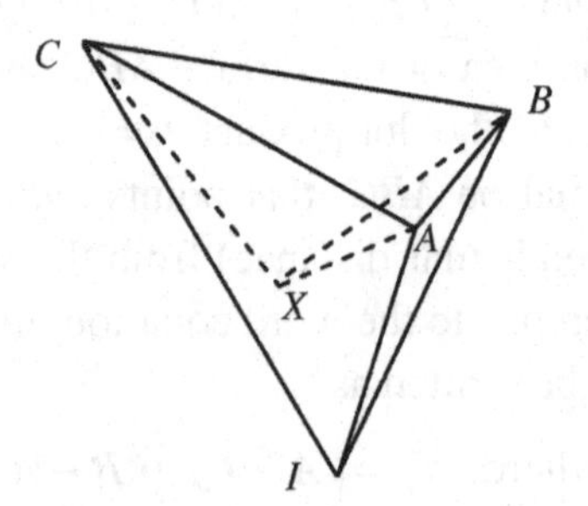

Fig. 10.5 Construction for volume shape function at node *I*

10.4 Volume Shape Functions

The concept of the area shape function can be generalized to three-dimensions using volume shape functions. For know nodal values of a quality $\phi_I, \phi_A, \phi_B, \phi_C$ the value of ϕ at any point with position vector $\boldsymbol{X}$ within the tetrahedron element of Fig. 10.5 can be approximated by the linear interpolation

$$\phi = N_I^V \phi_I + N_A^V \phi_A + N_B^V \phi_B + N_B^V \phi_B \tag{10.3}$$

where the N^V are the volume shape functions defined by

$$N_I^V = \frac{\text{Volume}_{\text{XABC}}}{\text{Volume}_{\text{IABC}}} = \frac{(\boldsymbol{A} - \boldsymbol{X}).((\boldsymbol{B} - \boldsymbol{A}) \times (\boldsymbol{C} - \boldsymbol{A}))}{T_I} \tag{10.4a}$$

$$N_A^V = \frac{(\boldsymbol{B} - \boldsymbol{X}) \cdot ((\boldsymbol{C} - \boldsymbol{B}) \times (\boldsymbol{I} - \boldsymbol{B}))}{T_A}$$

$$N_B^V = \frac{(\boldsymbol{C} - \boldsymbol{X}) \cdot ((\boldsymbol{I} - \boldsymbol{C}) \times (\boldsymbol{A} - \boldsymbol{C}))}{T_B} \tag{10.4b}$$

$$N_C^V = \frac{(\boldsymbol{I} - \boldsymbol{X}) \cdot ((\boldsymbol{A} - \boldsymbol{I}) \times (\boldsymbol{B} - \boldsymbol{I}))}{T_C}$$

Note by the construction of the N^V ,

$$\sum_{J=I,A,B,C} N_j^V = 1$$

throughout the element, and the shape function associated with the vertex $J(I, A, B \text{ or } C)$ vanishes on the opposite element surface. The derivatives of the shape function associated with vertex $\boldsymbol{I}$ (constant within the element) are given by

$$N_{Ix}^V = \frac{-\left[((\boldsymbol{B} - \boldsymbol{A}) \times (\boldsymbol{C} - \boldsymbol{A}))\right]_x}{T_I}$$

$$N_{Iy}^V = \frac{-\left[((\boldsymbol{B} - \boldsymbol{A}) \times (\boldsymbol{C} - \boldsymbol{A}))\right]_y}{T_I} \tag{10.5a}$$

$$N_{Iz}^V = \frac{-\left[((\boldsymbol{B} - \boldsymbol{A}) \times (\boldsymbol{C} - \boldsymbol{A}))\right]_z}{T_I}$$

where $[\boldsymbol{r}]_x$ indicates the x-component of the vector $\boldsymbol{r}$. Similar expressions for the derivatives of the three remaining shape functions can be readily obtained from (10.4b).

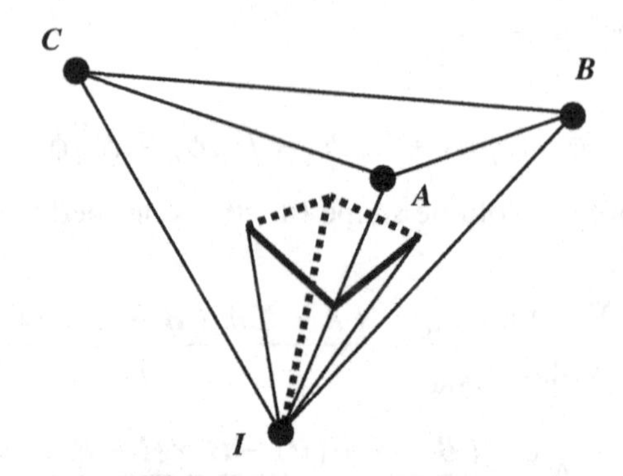

Fig. 10.6 A control volume face and section of control volume for node *i*

10.5 The Control Volume and Face

The control volume around point *i* in Fig. 10.2 is constructed by creating three(3) faces in each of the twenty-four (24) tetrahedron listed in Table 10.1. One of these faces is shown in Fig. 10.6. The face is created by defining the planar four-sided surface with vertices located at

the mid-point of the side *IA*
the mid point of surface *IAC*
the mid point of the tetrahedron *IABC*
the mid point of the surface *IAB*

Constructing the other two faces in the element to have vertices at the mid points of

side *IB*, surface *IAB*, tetrahedron *IABC*, surface *IBC*

and

side *IC*, surface *IBC*, tetrahedron *IABC*, surface *IAC*

will separate node *I* from the other nodes in the element. Similar constructions for all the tetrahedron in Fig. 10.2 with a common vertex on node I will completely enclose node I with a multi-faceted control volume consisting of seventy-six (76) piecewise continuous planar control volume faces. Important properties of a given face are its area its unit normal (pointing out of the control volume), and the volume of the tetrahedron element held between the face and the node I. Without loss of generality, in calculating the attributes of the control volume faces associated with node I we can place the origin at the point I. In this way the face in Fig. 10.6, will have vertices at

$$\tfrac{1}{2}\boldsymbol{A}, \tfrac{1}{3}(\boldsymbol{A}+\boldsymbol{B}), \tfrac{1}{4}(\boldsymbol{A}+\boldsymbol{B}+\boldsymbol{C}), \tfrac{1}{3}(\boldsymbol{A}+\boldsymbol{C})$$

see Fig. 10.7

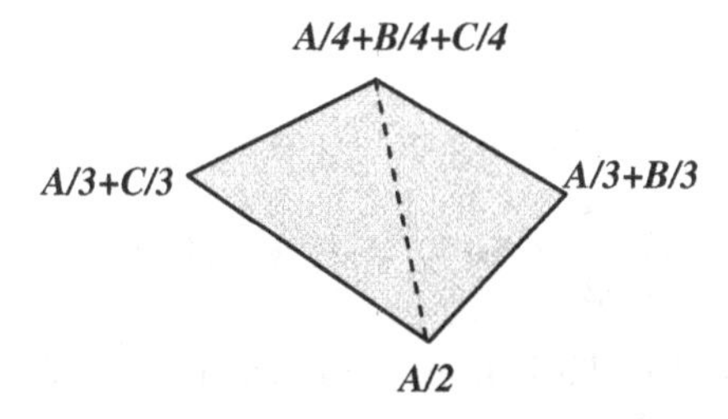

Fig. 10.7 Close up of control volume face of Fig 10.6 showing vertex vectors

With reference to the triangle on the right-hand side of the face in Fig. 10.7, the cross product

$$(\tfrac{1}{3}\boldsymbol{B}-\tfrac{1}{6}\boldsymbol{A})\times(\tfrac{1}{4}\boldsymbol{C}-\tfrac{1}{12}\boldsymbol{B}-\tfrac{1}{12}\boldsymbol{A})=\tfrac{1}{12}\boldsymbol{B}\times\boldsymbol{C}+\tfrac{1}{24}\boldsymbol{A}\times\boldsymbol{B}+\tfrac{1}{24}\boldsymbol{C}\times\boldsymbol{A} \quad (10.6)$$

Will give a vector in a normal direction to this surface with a magnitude equal to twice the area of the triangle. Likewise the cross product

$$(\tfrac{1}{4}\boldsymbol{C}-\tfrac{1}{12}\boldsymbol{B}-\tfrac{1}{12}\boldsymbol{A})\times(\tfrac{1}{6}\boldsymbol{A}-\tfrac{1}{3}\boldsymbol{C})=\tfrac{1}{12}\boldsymbol{B}\times\boldsymbol{C}+\tfrac{1}{24}\boldsymbol{A}\times\boldsymbol{B}+\tfrac{1}{24}\boldsymbol{C}\times\boldsymbol{A} \quad (10.7)$$

gives the equivalent vector for the left-hand triangular surface. Since both vectors in (10.6) and (10.7) are the same we can conclude that the construction of the control volume face provides a planar surface with an area normal product

$$(\mathbf{An})_A = \begin{cases} \left[\tfrac{1}{12}\mathbf{B}\times\mathbf{C} + \tfrac{1}{24}\mathbf{A}\times\mathbf{B} + \tfrac{1}{24}\mathbf{C}\times\mathbf{A}\right], & T_I > 0 \\ -\left[\tfrac{1}{12}\mathbf{B}\times\mathbf{C} + \tfrac{1}{24}\mathbf{A}\times\mathbf{B} + \tfrac{1}{24}\mathbf{C}\times\mathbf{A}\right], & T_I < 0 \end{cases} \tag{10.8a}$$

Pointing out of the ith control volume; the subscript $[]_A$ indicates that this face separates node I from node A. The equivalent area normal products for the other two node I control faces in the element of Fig 10.6 are

$$(\mathbf{An})_B = \begin{cases} \left[\tfrac{1}{24}\mathbf{B}\times\mathbf{C} + \tfrac{1}{24}\mathbf{A}\times\mathbf{B} + \tfrac{1}{12}\mathbf{C}\times\mathbf{A}\right], & T_I > 0 \\ -\left[\tfrac{1}{24}\mathbf{B}\times\mathbf{C} + \tfrac{1}{24}\mathbf{A}\times\mathbf{B} + \tfrac{1}{12}\mathbf{C}\times\mathbf{A}\right], & T_I < 0 \end{cases} \tag{10.8b}$$

$$(\mathbf{An})_C = \begin{cases} \left[\tfrac{1}{24}\mathbf{B}\times\mathbf{C} + \tfrac{1}{12}\mathbf{A}\times\mathbf{B} + \tfrac{1}{24}\mathbf{C}\times\mathbf{A}\right], & T_I > 0 \\ -\left[\tfrac{1}{24}\mathbf{B}\times\mathbf{C} + \tfrac{1}{12}\mathbf{A}\times\mathbf{B} + \tfrac{1}{24}\mathbf{C}\times\mathbf{A}\right], & T_I < 0 \end{cases} \tag{10.8c}$$

The volume held under the face of (10.6) is given by

$$V_A = \left|\frac{1}{3}\frac{\mathbf{A}}{2}\cdot(\mathbf{An})_A\right| = \left|\tfrac{1}{72}\mathbf{A}\cdot(\mathbf{B}\times\mathbf{C})\right| \tag{10.9}$$

which, on reference to (10.1a), is seen as 1/12th of the tetrahedron element volume. The other faces associated with node I hold identical volumes, hence the contribution to the ith control volume in Fig 10.2 from a given element connected to node i is 1/4th the volume of the element.

10.6 Approximation of Face Fluxes

The essential ingredient in constructing a CVFEM discrete equation for a given node is the approximation of diffusive and advective fluxes across the face of the control volume in terms of the nodal values of the element that contains the face in question.

10.6.1 Diffusive flux

The quantity per time entering the control volume across the volume face in Fig. 10.6 via diffusion is

$$Q_A^D = \kappa \nabla \phi \cdot (\, A\boldsymbol{n}\,)_A$$

Expanding in terms of the element nodal values

$$\begin{aligned} Q_A^D = \kappa(\, N_{Ix}^V \phi_I + N_{Ax}^V \phi_A + N_{Bx}^V \phi_B + N_{Cx}^V \phi_C\,)(\, An_x\,)_A + \\ \kappa(\, N_{Iy}^V \phi_I + N_{Ay}^V \phi_A + N_{By}^V \phi_B + N_{Cy}^V \phi_C\,)(\, An_y\,)_A + \\ \kappa(\, N_{Iz}^V \phi_I + N_{Az}^V \phi_A + N_{Bz}^V \phi_B + N_{Cz}^V \phi_C\,)(\, An_z\,)_A \end{aligned} \tag{10.10}$$

Gathering terms will identify the contributions to the respective nodal coefficients in the CVFEM equation for node I.

10.6.2 Advective flux

For a given velocity $\boldsymbol{v}$ the discharge across the control face in Fig. 10.6 is

$$Q_A = \boldsymbol{v} \cdot (\, A\boldsymbol{n}\,)_A$$

Using upwinding the quantity per time entering the volume by advection across the face can be approximated as

$$Q_A^U = \begin{cases} -Q_A \phi_I, & Q_A > 0 \\ -Q_A \phi_A, & Q_A < 0 \end{cases} \tag{10.11}$$

and then used to update the appropriate nodal coefficients in the CVFEM equation for node i.

10.7 Summary

The reader should recognize that, in the 3-D case, the basic principal of arriving at a discrete equation by balancing the flow of a quantity into the control volume is identical to the two-dimensional case fully described in Chapter 5. Above all the information has been provided to evaluate the coefficients on the right-hand sides of (10.10) and (10.11). These form the basic building block of a 3-D CVFEM solution. From this point on, the additional requirements to arrive at a working 3-D code, although involved, are essentially no more than a labeling and bookkeeping exercise.

Appendix A

A Meshing Code

This appendix provides a MATLAB code (MESH.m) for creating a mesh of linear triangular elements. The data structure generated matches the unstructured data structure presented in Chapter 3 and used in the CVFEM developments in Chapter 5. Note, however, that this mesh is generated by initially assuming that the nodes points are laid out on a structured grid of rows and columns. On the grid the morphology of the region of support i always have the morphology shown in Fig. A1. Further, the application of this code is limited to domains with 4 distinct sides that can be covered by a grid of rows and columns. The obvious domains are rectangles and squares, but with some imagination it is possible to find and mesh other domains that have a similar morphology; two important examples, illustrated in Fig. A.2, are an annulus (a ¼ section is shown) and a plate with a hole (an 1/8 section is shown).

Please note:

(1) The code MESH is not an attempt to win the MATLAB program of the year. Indeed, the author has purposely attempted to write the most basic code, avoiding, where possible, any build in MATLAB functions and efficiencies. Hence the resulting code may not be efficient but it is understandable and will allow for the translation to other coding languages.

(2) The code MESH is provided so that readers can get a rapid start on programming and using the CVFEM. In more general application the experienced reader is encouraged to use an alternative un-structured mesh generator followed by an appropriate "translation" to generate the data structure outlined in Chapter 3.

(3) The mesh generated is given in Fig. B1. (appendix B).

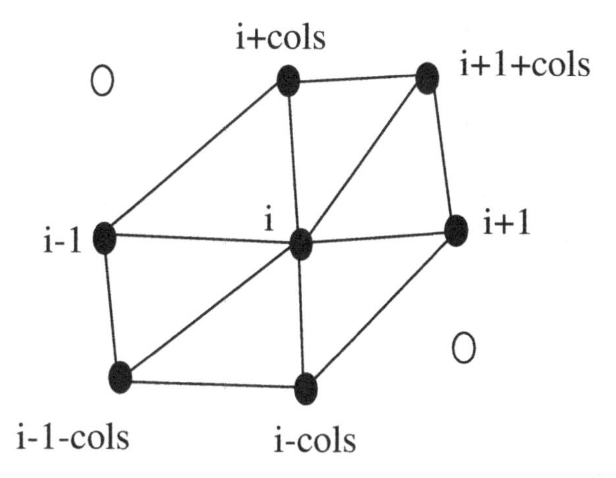

Fig. A1 The assumed morphology of the region of support in the program MESH

```
%MESH
%V.R. Voller
%volle001@umn.edu

%This program creates a triangular mesh and its data structure
%From a row-col(umn) structure laid out on a 4-sided domain
%The 1/4 annulus domain( see Fig A2) is programmed here

clear all
rows=21;
cols=21;

n=rows*cols; %Total domain nodes

% Global nodes points stored in vectors x and y
% Current settings for 1/4 annulus

rin=1; %inner radius
```

```
rot=2; %outer radius
for k=1 : rows
   for j=1 : cols
      theta=2*atan(1)*(k-1)/(rows-1);        %angle
      radius=rin+(rot-rin)*(j-1)/(cols-1);   %radial pos.

      i=j+(k-1)*cols;                   %converts row-column to single index

      x(i)=radius*cos(theta);
      y(i)=radius*sin(theta);
   end
end

%zero out data structure

%support
for k=1:rows
   for j=1:cols
      i=j+(k-1)*cols;
      n_s(i)=0;      %Nodes in support
      for ii=1:12
         S(i,ii)=0;  %Global node number of ii_th node in support i
      end
   end
end

%boundary
n_seg=4;               %boundary segments
n_b(1)=cols;
n_b(2)=rows;        %outter radius
n_b(3)=cols;
n_b(4)=rows;        %inner radius

for k=1:n_seg
   for j=1:n_b(k)
     B(k,j)=0;      %Global node number of j_th node on i_th boundary
   end
end
```

```
%SUPPORT DATA ------------

%Assumed Morphology in Fig. A1

% Support for internal points
for k=2:rows-1
   for j=2:cols-1

      i=j+(k-1)*cols;
      n_s(i)=6;

      S(i,1)=i+1;      %assumed morphology  for  support see  Fig. A1
      S(i,2)=i+1+cols;
      S(i,3)=i+cols;
      S(i,4)=i-1;
      S(i,5)=i-1-cols;
      S(i,6)=i-cols;
      S(i,7)=i+1;     %last node is same as first node at internal nodes
   end
end

%horizonal boundary (y=0) Non-corner points
   for j=2:cols-1
      i=j;
      n_s(i)=4;
      S(i,1)=i+1;         %  arranged so that points are--
      S(i,2)=i+1+cols;  %  --contiguous in counter clockwise count
      S(i,3)=i+cols;
      S(i,4)=i-1;
      S(i,5)=0;           % last point zero: a boundary node flag
   end


%outter (rot) curve boundary Non-corner points
  for k=2:rows-1
      i=cols+(k-1)*cols;
      n_s(i)=4;
      S(i,1)=i+cols;
      S(i,2)=i-1;
```

```
        S(i,3)=i-1-cols;
        S(i,4)=i-cols;
        S(i,5)=0;
    end

% vertical boundary (x=0) Non-corner points
    for j=2:cols-1
        i=j+(rows-1)*cols;
        n_s(i)=4;
        S(i,1)=i-1;
        S(i,2)=i-1-cols;
        S(i,3)=i-cols;
        S(i,4)=i+1;
        S(i,5)=0;
    end

%inner (rin) curve boundary Non-corner points
    for k=2:rows-1
        i=1+(k-1)*cols;
        n_s(i)=4;
        S(i,1)=i-cols;
        S(i,2)=i+1;
        S(i,3)=i+1+cols;
        S(i,4)=i+cols;
        S(i,5)=0;
    end

%Corners
  % x=0, y = rin
  i=1;
  n_s(i)=3;
  S(i,1)=i+1;
  S(i,2)=i+1+cols;
  S(i,3)=i+cols;
  S(i,4)=0;

  % y = 0 x = rot
  i=cols;
  n_s(i)=2;
```

```
  S(i,1)=i+cols;
  S(i,2)=i-1;
  S(i,3)=0;

  % x =0 y = rot
  i=rows*cols;
  n_s(i)=3;
  S(i,1)=i-1;
  S(i,2)=i-1-cols;
  S(i,3)=i-cols;
  S(i,4)=0;

  % x=0 y = rin
  i=1+(rows-1)*cols;
  n_s(i)=2;
  S(i,1)=i-cols;
  S(i,2)=i+1;
  S(i,3)=0;

%END SUPPORT DATA-----------

% BOUNDARY DATA------------
%NOTE counter-clockwise numbering and boundary segment index

  % y=0 boundary (segment 1)
  for i=1:n_b(1)
     B(1,i)=i;
  end

  %  r=rout boundary (segment 2)
  for i=1:n_b(2)
     B(2,i)=i*cols;
  end

  % x =0 boundary (segment 3)
  for i=1:n_b(3)
     B(3,i)=rows*cols+1-i;
  end
```

```
 % r=rin boundary (segment 4)
 for i=1:n_b(4)
    B(4,i)=1+rows*cols-i*cols;
 end
%END BOUNDARY DATA-----------------

%Print Out mesh from supports (MATLAB functions used)
 hold on
 axis equal
  for i=1:n
    clear xtemp;
    clear ytemp;
    ytemp(1)=y(i);
    xtemp(1)=x(i);
      for j=1:n_s(i)
          xtemp(2)=x(S(i,j));
          ytemp(2)=y(S(i,j));
          plot(xtemp,ytemp,'k');
    end
  end
  %hold off
```

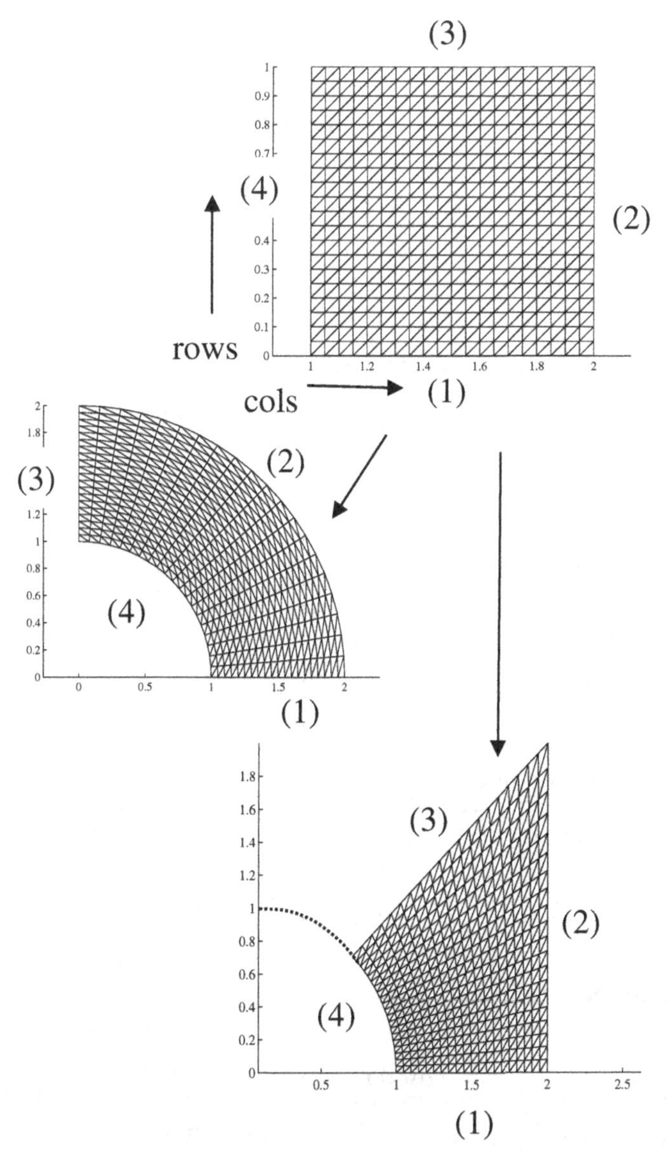

Fig. A2 Possible domain morphologies that can be meshed with MESH. The case in the program given is the ¼ annulus

Appendix B

A CVFEM Code

This appendix provides a MATLAB code (CVFEM.m) that implements a CVFEM solution for the steady state advection diffusion in an annulus. The full problem definition is given in §7.7.2. Here the governing equation is written in two-dimensional Cartesian form

$$\frac{\partial v_x \phi}{\partial x} + \frac{\partial v_y \phi}{\partial y} - \frac{\partial}{\partial x}\left(\kappa \frac{\partial \phi}{\partial x}\right) - \frac{\partial}{\partial y}\left(\kappa \frac{\partial \phi}{\partial y}\right) = 0,$$

in

$$x \geq 0,\ y \geq 0,\ 1 \leq \sqrt{x^2 + y^2} \leq 2$$

With boundary conditions

$$\phi = 1,\ on\ \sqrt{x^2 + y^2} = 1,\ \phi = 0,\ on\ \sqrt{x^2 + y^2} = 2$$

$$\frac{\partial \phi}{\partial y} = 0,\quad on\quad y = 0,\qquad \frac{\partial \phi}{\partial x} = 0,\quad on\quad x = 0$$

and parameter definitions

$$\kappa = \frac{1}{\sqrt{x^2 + y^2}},\ v_x = \frac{cos(\theta)}{\sqrt{x^2 + y^2}},\ v_y = \frac{sin(\theta)}{\sqrt{x^2 + y^2}},\ \theta = tan^{-1}(y/x)$$

This equation is solved in the code CVFEM using the mesh and data structure generated by the program MESH in Appendix A. The domain and the mesh used is shown in Fig. B1.

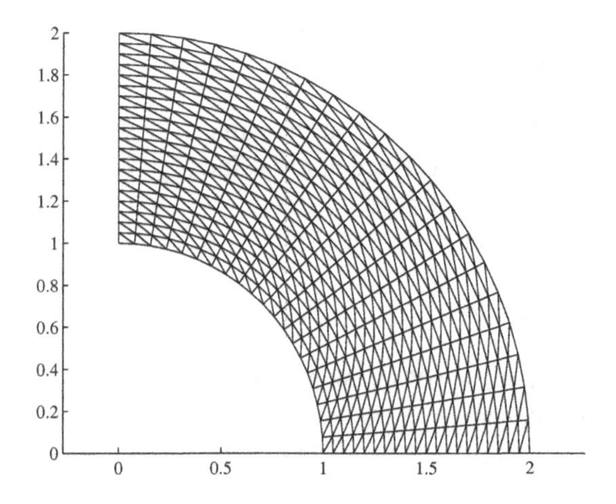

Fig. B1 Mesh used in program CVFEM

As in Appendix A , please note:

(1) The code CVFEM is not an attempt to win the MATLAB program of the year. Indeed, the author has purposely attempted to write the most basic code, avoiding, where possible, any build in MATLAB functions and efficiencies. Hence the resulting code may not be efficient but it is understandable and will allow for the translation to other coding languages.

(2) The code CVFEM is provided so that readers can get a rapid start on programming and using the CVFEM. Although it is restricted to the case of steady state advection-diffusion it is sufficient to illustrate the essential layout of a two-dimensional CVFEM code and, with some effort from the reader, provide a starting point for the key CVFEM programming for all the problems presented in this volume.

```
%CVFEM
%V.R. Voller
%volle001@umn.edu

%current set up steady-state advection-diffusion in a 1/4 annulus.
%Mesh and data structure obtained from program MESH.m

%The inner (r = 1) radius is fixed at the value 1
%The outer radius (r = 2) is fixed at the value 0
%The radial velocity is v_r = 1/r
%The diffusivity is kappa=1/r

%radial value and angle
for i=1:n
   radius(i)=sqrt(x(i)^2+y(i)^2);  % radial position
   theta(i)=atan2(y(i),x(i));      % angle
end

% x-y Velocity field
for i=1:n
   vx(i)=cos(theta(i))/radius(i);
   vy(i)=sin(theta(i))/radius(i);
end

% SET CVFEM COEFICIENTS-----------------

%zero out coefficient

for i=1:n  %loop on nodes

  %zero out coefficient
  ap(i)=0.0;     %node point coefficient
  V(i)=0.0;      %volume of CV

  for j=1:n_s(i)+1
    a(i,j)=0.0;  %neighbor coefficient
  end
```

```
% Set loopcount

% At internal nodes--recognized by S(i,n_s(i)+1)<>0
% the nodes in the support match the elements in the support
% Hence loop over every node in the node i support: loopcount=n_s(i)

% At boundary nodes--recognized by  S(i,n_s(i)+1)=0
% the nodes in the support are one larger than the elements in the support
% Hence only loop over the first n_s-1 nodes: loopcount=n_s(i)


  loopcount=n_s(i);
  if S(i,n_s(i)+1)==0
      loopcount=n_s(i)-1;
  end


% Pick out each element in turn from i_th support

  for j=1:loopcount

    k1=S(i,j);      %The numbers of the nodes in the j_th elemnt are
    k2=S(i,j+1);    %  i, k1=S(i,J), and k2=S(i,j+1)

    xL(1) = x(i);   %xL=local x-coordinate of a node within a element
    xL(2) = x(k1);
    xL(3) = x(k2);
    yL(1) = y(i);   %yL=local y-coordinate of a node within a element
    yL(2) = y(k1);
    yL(3) = y(k2);

%   nodal diffusivity (kappa)
    diff(1)=1/sqrt(xL(1)^2+yL(1)^2);
    diff(2)=1/sqrt(xL(2)^2+yL(2)^2);
    diff(3)=1/sqrt(xL(3)^2+yL(3)^2);
```

```
% nodal velocities
  vxL(1)=vx(i);
  vxL(2)=vx(k1);
  vxL(3)=vx(k2);
  vyL(1)=vy(i);
  vyL(2)=vy(k1);
  vyL(3)=vy(k2);

% Vele= element-volume
  Vele=(xL(2)*yL(3)-xL(3)*yL(2)-xL(1)*yL(3)+xL(1)*yL(2)...
        +yL(1)*xL(3)-yL(1)*xL(2))/2;


  V(i)=V(i)+Vele/3; %Contribution to control volume

  Nx(1)= (yL(2)-yL(3))/(2*Vele); %derivative of shape function wrt x
  Nx(2)= (yL(3)-yL(1))/(2*Vele);
  Nx(3)= (yL(1)-yL(2))/(2*Vele);
  Ny(1)=-(xL(2)-xL(3))/(2*Vele); %derivative of shape function  wrt y
  Ny(2)=-(xL(3)-xL(1))/(2*Vele);
  Ny(3)=-(xL(1)-xL(2))/(2*Vele);

% Face 1--values of diffusivity and velocity
  diff_f=(5*diff(1)+5*diff(2)+2*diff(3))/12;
  vxf=  (5*vxL(1)+5*vxL(2)+2*vxL(3))/12;
  vyf=  (5*vyL(1)+5*vyL(2)+2*vyL(3))/12;

% Face 1--face area normal components
  delx= [(xL(1)+xL(2)+xL(3))/3]-[(xL(1)+xL(2))/2];
  dely= [(yL(1)+yL(2)+yL(3))/3]-[(yL(1)+yL(2))/2];

  qf=vxf*dely-vyf*delx; %Face 1 volume flux

% Face 1--Diffusion Coefficients
  ap(i) =   ap(i)      + diff_f*(- Nx(1) * dely + Ny(1) * delx);
  a(i,j) =  a(i,j)     + diff_f*(  Nx(2) * dely - Ny(2) * delx);
  a(i,j+1) = a(i,j+1) + diff_f*(  Nx(3) * dely - Ny(3) * delx);
```

```
% Face 1--Advection Coefficients
    ap(i)= ap(i) +max(qf,0);
    a(i,j)=a(i,j) +max(-qf,0);

% Face 2--values of diffusivity and velocity
    diff_f=(5*diff(1)+2*diff(2)+5*diff(3))/12;
    vxf=( 5*vxL(1)+2*vxL(2)+5*vxL(3))/12;
    vyf=( 5*vyL(1)+2*vyL(2)+5*vyL(3))/12;

% Face 2--face area normal components
    delx= [(xL(1)+xL(3))/2]-[(xL(1)+xL(2)+xL(3))/3];
    dely= [(yL(1)+yL(3))/2]-[(yL(1)+yL(2)+yL(3))/3];

    qf=vxf*dely-vyf*delx;% Face 1 volume flux

% Face 2--Diffusion Coefficients
    ap(i) =   ap(i)     + diff_f*(- Nx(1) * dely + Ny(1) * delx);
    a(i,j) =  a(i,j)    + diff_f*( Nx(2) * dely - Ny(2) * delx);
    a(i,j+1) = a(i,j+1) + diff_f*( Nx(3) * dely - Ny(3) * delx);

% Face 2--Advection Coefficients
   ap(i)=   ap(i)     + max(qf,0);
   a(i,j+1) = a(i,j+1) + max(-qf,0);

  end %end loop on nodes/elements in support

% For internal nodes LAST node in region of support is also the FIRST
% Hence coefficient for last node needs to be added to first node
% and then removed

 a(i,1)=a(i,1)+a(i,n_s(i)+1);
 a(i,n_s(i)+1)=0;

end   %end loop on nodes in domain

% END CVFEM COEFFICIENTS-----------
```

```
%SET BOUNDARY

for i=1:n  %Default values
   BC(i)=0;  %coefficient part
   BB(i)=0;  %constant part
end


% Set value = 0 on r = r_outer boundary segment 2
for k=1:n_b(2)
      BC(B(2,k))=1e18;
end

%set value =1 on r = r_inner boundary segment 4
for k=1:n_b(4)
      BC(B(4,k))=1e18;
      BB(B(4,k))=1e18;
end

%-----------------------------

%SET SOURCE  (No source in current problem)
for i=1:n
   QC(i)=0;
   QB(i)=0;
end

%------------------------
```

```
%SOLVER

%Very Crude point iteration Solver
%Caution There is no formal convergence check
%Increase the size of the iteration loop to ensure convergence

%Defualt initial guess

for i=1:n
   phi(i)=0;   %Dependent variable
end


for iter=1:1500          %Choose sufficient iteration for convergence

  for i= 1:n
     RHS=BB(i)+QB(i);
     for j=1:n_s(i)
       RHS=RHS+a(i,j)*phi(S(i,j));
     end
     phi(i)=RHS/(ap(i)+BC(i)+QC(i));
  end

end

%END SOLVRT

%store phi values on x-axis
pout=0;
for i=1:n
   if y(i)<1e-5
       pout=pout+1;
       xout(pout)=x(i);
       phiout(pout)=phi(i);
   end
end
```

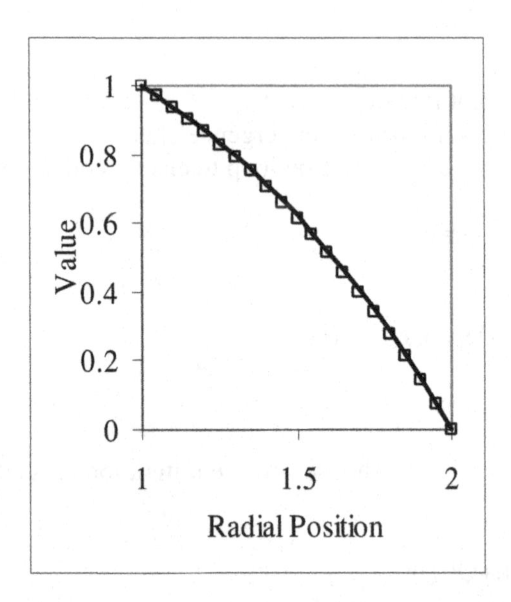

Fig. B2 Results from CVFEM (open symbols) compared with analytical solution (line)

The solution from this code, the profile of ϕ along the x-axis is compared with the analytical solution

$$\phi = \frac{e^r - e^2}{e - e^2}, \quad r = \sqrt{x^2 + y^2}$$

In Fig. B2. In addition to provide a detailed comparison some point values are given in Table B1.

Table B1 Comparison of CVFEM and analytical predictions

x	CVFEM	Analytical
1	1	1
1.2	0.8667	0.871149
1.4	0.706	0.713769
1.6	0.5119	0.521546
1.8	0.2783	0.286764
2	0	0

Bibliography

Abramowitz, M. and Stegun I. A. (1970). *Handbook of Mathematical Functions*, Dover, New York.

Baliga, B.R., and Atabaki, N. (2006). Control-volume-based finite-difference and finite-element methods, *Handbook of Numerical Heat Transfer (eds Minkowycz, W .J., Sparrow, E. M. and Murthy, J. Y)*, Wiley, Hoboken.

Baliga, B.R., and Patankar, S.V. (1980). A new finite element formulation for convection-diffusion problems, *Numer. Heat Transfer,* 3, pp. 393-409.

Batchelor, G.K. (1970). *An Introduction to Fluid Dynamics*, Cambridge University Press.

Barber, J.R. (2003). *Elasticity* (2nd *edn.*), Kluwer Academic Publishers, Dordrecht / Boston.

Crank, J. and Nicolson, P. (1947). A practical method for numerical evaluation of solutions of partial differential equations of the heat conduction type, *Proc. Cambridge Phil. Soc.* 43, pp. 50–64.

Ghia U, Ghia KN, Shin CT. (1982). High-Re solutions for incompressible flow using the Navier-Stokes equations and a multigrid method, *Journal of Computational Physics,* 48, pp. 387-411.

Hsieh, P.A. (1986). A new formula for the analytical solution of the radial dispersion problem, *Water Resources Research*, 22, pp. 1597-1605.

Kikuchi, N. (1986). *Finite Element Methods in Mechanics*, Cambridge University Press.

Fryer, Y.D., Bailey, C. Cross, M. and Lai, C.-H. (1991). A control-volume procedure for solving the elastic stress-strain equations on an unstructured mesh, *Applied Math. Model.,* 15, pp. 639-645.

Patankar, S. V. (1980). *Numerical Heat Transfer and Fluid Flow*, Hemisphere, Washington.

Pepper, D. W. (2006). Meshless methods, *Handbook of Numerical Heat Transfer (eds Minkowycz, W .J., Sparrow, E. M, and Murthy, J. Y),* Wiley, Hoboken.

Smith G.D. (1985). *Numerical Solution of Partial Differential Equations: Finite Difference Methods* (3rd *edn.*), Oxford University Press.

Winslow, A.M. (1966). Numerical solution of the quasilinear Poisson equation in a nonuniform triangular mesh, *J. Comp. Phys.,* 1, pp. 149-172.

Woodfield, P. L., Suzuki, K., and Nakabe, K. (2004). A simple strategy for constructing bounded convection schemes for unstructured grids, *International Journal for Numerical Methods in Fluids*, 46, pp. 1007-1024.

Zienkiewicz, O. C. and Taylor, R. L. (1989). *The Finite Element Method* (*4th edn.*), McGraw-Hill, London.

Index